# COOPERATIVE GAMES, SOLUTIONS AND APPLICATIONS

**THEORY AND DECISION LIBRARY**

General Editors: W. Leinfellner and G. Eberlein

Series A: Philosophy and Methodology of the Social Sciences

Editors: W. Leinfellner (Technical University of Vienna)
G. Eberlein (Technical University of Munich)

Series B: Mathematical and Statistical Methods

Editor: H. Skala (University of Paderborn)

Series C: Game Theory, Mathematical Programming and Operations Research

Editor: S. H. Tijs (University of Nijmegen)

Series D: System Theory, Knowledge Engineering and Problem Solving

Editor: W. Janko (University of Vienna)

---

## SERIES C: **GAME THEORY, MATHEMATICAL PROGRAMMING AND OPERATIONS RESEARCH**

Editor: S. H. Tijs (University of Nijmegen)



**Scope**

Particular attention is paid in this series to game theory and operations research, their formal aspects and their applications to economic, political and social sciences as well as to sociobiology.
It will encourage high standards in the application of game theoretical methods to individual and social decision making.

For list of titles in this series, see final page.

# COOPERATIVE GAMES, SOLUTIONS AND APPLICATIONS

*by*

THEO DRIESSEN
*Assistant Professor, Faculty of Applied Mathematics, University of Twente, The Netherlands*

KLUWER ACADEMIC PUBLISHERS
DORDRECHT / BOSTON / LONDON

Library of Congress Cataloging in Publication Data

CIP

Driessen, Theo, 1957-
Cooperative games, solutions, and applications.

(Theory and decision library. Series C, Game theory, mathematical programming, and operations research)
Bibliography: p.
Includes indexes.
1. Game theory. I. Title. II. Series.
QA269.D75 1988 519.3 88-6809

ISBN 90-277-2729-5

Published by Kluwer Academic Publishers,
P.O. Box 17, 3300 AA Dordrecht, The Netherlands.

Kluwer Academic Publishers incorporates
the publishing programmes of
D. Reidel, Martinus Nijhoff, Dr W. Junk and MTP Press.

Sold and distributed in the U.S.A. and Canada
by Kluwer Academic Publishers,
101 Philip Drive, Norwell, MA 02061, U.S.A.

In all other countries, sold and distributed
by Kluwer Academic Publishers Group,
P.O. Box 322, 3300 AH Dordrecht, The Netherlands.



Printed in The Netherlands

# PREFACE

The study of the theory of games was started in Von Neumann (1928), but the development of the theory of games was accelerated after the publication of the classical book "Theory of games and economic behavior" by Von Neumann and Morgenstern (1944).

As an initial step, the theory of games aims to put situations of conflict and cooperation into mathematical models. In the second and final step, the resulting models are analysed on the basis of equitable and mathematical reasonings. The conflict and/or cooperative situation in question is generally due to the interaction between two or more individuals (players). Their interaction may lead up to several potential payoffs over which each player has his own preferences. Any player attempts to achieve his largest possible payoff, but the other players may also exert their influence on the realization of some potential payoff.

As already mentioned, the theory of games consists of two parts, a modelling part and a solution part. Concerning the modelling part, the mathematical models of conflict and cooperative situations are described. The description of the models includes the rules, the strategy space of any player, potential payoffs to the players, the preferences of each player over the set of all potential payoffs, etc. According to the rules, it is either permitted or forbidden that the players communicate with one another in order to make binding agreements regarding their mutual actions. The cooperative (noncooperative respectively) game theory deals with the mathematical models of situations in which cooperation between the players is permitted (forbidden). The fundamental modelling approach to the theory of games was presented in Von Neumann and Morgenstern (1944). The mathematical theory of games is founded, more or less, on one of the following three models:

- the extensive (or tree) form
- the normal (or strategic) form and
- the characteristic function (or coalitional) form.

Concerning the solution part, the resulting payoffs to the players are determined according to certain solution concepts. Here any solution concept is based on a specific interpretation of the fairness of some potential payoffs. The relevant criteria of a fair payoff are many and varied. Thus, it is not surprising that various solution concepts for the three main models have been proposed since 1944.

In this work the focus is on the theory of cooperative games in characteristic function form. The modelling theory of these games is treated in the introductory Chapter I and a thorough treatment of the standard solution theory of the games is given in Chapter II. There are two main features which set this work apart. As the first feature, the theory of an additional solution concept, called the $\tau$-value, is developed in Chapter III. As the second feature, the notion of k-convexity for cooperative games in characteristic function form is introduced and studied in Chapter VII. The notion of k-convexity is closely related to the well-known convexity notion and therefore, the theory of convex games is treated in Chapter V. The remaining Chapters IV and VI illustrate two significant applications of cooperative game theory. A game theoretic approach to the cost allocation problem as well as the bankruptcy problem is presented in such a way that cost allocation methods and division rules for bankruptcy problems are compared with game theoretic solution concepts such as the $\tau$-value.

During the eighties, many new developments took place within the dynamic field of game theory. This work covers four recent developments regarding cooperative games in characteristic function form: the $\tau$-value, the k-convexity and a game theoretic approach to the bankruptcy problem as well as the cost allocation problem (of which the research was already started in the beginning of the seventies). The theory of both the $\tau$-value and the k-convex games is an outgrowth of the game theoretic research performed in The Netherlands. Besides, twelve Dutch game theoreticians have compiled a book consisting of surveys on the state of the art of their specific topics within or related to the field of (non)cooperative game theory, supplemented with recent developments. Their book is entitled *Surveys in Game Theory and Related Topics* (Eds. H.J.M. Peters and O.J. Vrieze), CWI Tract 39, Centre for Mathematics and Computer Science, Amsterdam, 1987.

The purpose of this work is to present a comprehensive treatment of two main features ($\tau$-value, k-convexity) as well as the standard/recent cooperative game theory which is related, more or less, to the two main features. The relevant mathematical text is written as simple as possible for students with a basic knowledge of mathematics. That is, the text should be understandable to students who are familiar with the elementary way of mathematical reasoning. No list of exercises is included. Throughout each chapter, it is left to the reader to check various relevant facts which are verifiable in an almost straightforward way.

Finally, the author wants to thank the Dutch and some foreign colleagues for their stimulating discussions and their (in)direct contributions to this work. Further, the author is grateful to the Faculty of Applied Mathematics, University of Twente, Enschede, for granting secretarial facilities. Special thanks are due to Anja Meulenbelt for both her excellent typewriting and her persistent devotion. The author accepts full responsibility for the errors in this work.

T.S.H. Driessen

TABLE OF CONTENTS

## NOTATION

| | denotes |
|---|---|
| $\mathbb{N} := \{1,2,3,\ldots\}$ | the set of natural numbers |
| $\mathbb{R}$ | the set of real numbers |
| $\emptyset$ | the empty set |
| dim | dimension |
| max | maximum |
| min | minimum |

Let $n \in \mathbb{N}$ and $N := \{1,2,\ldots,n\}$. Then

| | denotes |
|---|---|
| $2^N := \{S \mid S \subset N\}$ | the set of subsets of N |
| $\mathbb{R}^N$ | the set of real-valued functions on N |
| $\mathbb{R}^n$ | the space of n-tuples of real numbers |
| $e^i \in \mathbb{R}^n$ $(i \in N)$ | the i-th unit vector in $\mathbb{R}^n$ defined by $e^i_i := 1$ and $e^i_j := 0$ for all $j \in N-\{i\}$ |
| $e^S \in \mathbb{R}^n$ $(S \subset N)$ | the vector in $\mathbb{R}^n$ defined by $e^S_i := 1$ if $i \in S$; $= 0$ if $i \in N-S$ |

$\mathbb{R}^n_+ := \{(x_1,x_2,\ldots,x_n) \in \mathbb{R}^n \mid x_i \geq 0 \text{ for all } i \in N\}$

$1_n := (1, 1, \ldots, 1) \in \mathbb{R}^n_+$

$\Theta^n := \{\theta \mid \theta\colon N \to N, \theta \text{ is a permutation}\}$

| | denotes the class of |
|---|---|
| $B^n$ | balanced n-person games |
| $C^n$ | convex n-person games |
| $C^n_k$ | k-convex n-person games ($k \in \mathbb{N}$) |
| $CG^n$ | n-person cost games |
| $G^n$ | cooperative n-person games (with player set N) |
| $I^n$ | n-person games with a nonempty imputation set |
| $M^n_0$ | zero-monotonic n-person games |
| $QB^n$ | quasibalanced n-person games |
| $S^n$ | simple n-person games |
| $SC^n$ | semiconvex n-person games |

Let $x = (x_1,x_2,\ldots,x_n) \in \mathbb{R}^n$, $y = (y_1,y_2,\ldots,y_n) \in \mathbb{R}^n$. Then the vector inequality $x \geq y$ holds if $x_i \geq y_i$ for all $i \in N$, the vector equality $x = y$ holds if $x_i = y_i$ for all $i \in N$.

A real number $\alpha \in \mathbb{R}$ is said to be
positive if $\alpha > 0$, nonpositive if $\alpha \leq 0$,
negative if $\alpha < 0$, nonnegative if $\alpha \geq 0$.

A function $f: \mathbb{R} \to \mathbb{R}$ is said to be
increasing if $f(x) > f(y)$ for all $x, y \in \mathbb{R}$ with $x > y$,
decreasing if $f(x) < f(y)$ for all $x, y \in \mathbb{R}$ with $x > y$,
nonincreasing if $f(x) \leq f(y)$ for all $x, y \in \mathbb{R}$ with $x > y$,
nondecreasing if $f(x) \geq f(y)$ for all $x, y \in \mathbb{R}$ with $x > y$.

Let $X$ be a linear space over $\mathbb{R}$ and $Y \subset X$, $Y \neq \emptyset$. Then
conv $Y$ denotes the convex hull of the set $Y$ and
ext $Y$ denotes the set of extreme points of $Y$ in case $Y$ is a convex set.
A well-known theorem of Krein - Milman yields the equality $Y = \text{conv}(\text{ext } Y)$ whenever $Y$ is a nonempty compact convex subset of $\mathbb{R}^n$.

Corollaries, definitions, examples, lemmas, propositions and theorems within one chapter are regarded to be of the same type and are numbered successively on a per-section basis. The seven chapters are numbered by Roman numerals. On reference to a corollary et al. listed in a former or subsequent chapter, the Roman numerals are needed, whereas similar references within the same chapter are mentioned without Roman numerals. For instance, Theorem V.3.7 refers to Theorem 7 of Section 3 of Chapter V.
Figures and tables are regarded to be of a second type and are numbered successively on a per-chapter basis. E.g., Figure 6.1, Tables 6.2–6.3 and Figures 6.4–6.5 denote all the figures and tables listed in Chapter VI. In addition, various important formulas within one chapter are numbered successively by using the notation (•.•). That is, (7.24) refers to the formula 24 of Chapter VII.
Finally, □ denotes the end of a proof.

# CHAPTER I

# COOPERATIVE GAMES AND EXAMPLES

## 1. The Tennessee Valley Authority (Ransmeier, 1942)

The Tennessee River is a stream situated in the south-eastern United States of America. The so-called Muscle Shoals, a 37-mile stretch of the Tennessee River with a fall of 133 feet, was extremely suitable to the generation of power. Due to the first World War, the United States carried out a program of domestic air-nitrate production for munitions. Because the air-nitrate production required vast supplies of electric power, Muscle Shoals was selected in 1917 for federal development during the World War as a great national defense hydroelectric - air nitrate centre. But the two nitrate plants, the hydroelectric dam and the power plant were not completed in time to contribute to the war munitions effort. For instance, the Wilson Dam was completed in 1925 after many interruptions. As early as 1919, when the unifying issue of national defense was removed, the public power - private power controversy developed with respect to an appropriate use or disposition of the federal Muscle Shoals properties. By 1933 this controversy had led up to an almost hopeless impasse.
Concerning the social-economic milieu of the Tennessee Valley region, the basis problem was the poverty of its predominantly rural population. Recognizing the problems and the potentialities of the Tennessee Valley, Congress adopted an act to establish the <u>Tennessee Valley Authority</u> (TVA), a regional agency of the federal government of 1933 which was dedicated to a comprehensive valley-wide program to improve economic opportunity in the valley region. The TVA did this through activities along four major lines:

(i) soil conservation, e.g., by means of production of concentrated phosphatic fertilizers in the converted nitrate plants (in which way the terms of Section 124 of the National Defense Act of 1916 were met);

(ii) technical research and industrial development, e.g., development and production of new farm machinery;

(iii) water control: navigation, flood control and development of power;
(iv) power marketing, e.g., payable electricity.

In executing its whole program, the TVA has accomplished substantial results, mainly because of its character as a decentralized, multiple purpose federal agency.

Of the three purposes of water control the one which has been carried out on the greatest scale for the production of revenue is power. In view of the public power - private power controversy of the past decade and the fact that the cost-revenue relation for the purpose of power in the TVA's program would have a major impact on the cost of publicly produced power as compared with privately produced power, it is not surprising that the TVA had also to deal with the principle of cost allocation. As a matter of fact, Section 14 (the allocation section) of the TVA Act of 1933 required the Authority to prepare allocations of the cost of the Wilson Dam and of such additional reservoir projects as TVA might construct. Allocation of costs was required among five objectives:

(1) navigation;
(2) flood control;
(3) development of power; and to a less extent
(4) national defense and
(5) fertilizer production.

Notably absent in the allocation section of the TVA Act of 1933 was the condition that the sum of the costs allocated to the various objectives should be equal to the total cost of the project. Nevertheless, most of the allocation methods which were studied under the direction of the Authority, met this condition. The problem was to choose an allocation which is equitable and defensible. Ransmeier (1942) listed five "preliminary criteria of a satisfactory allocation". His first preliminary criterion is stated as follows (cf. page 220):

"The method should have a reasonable logical basis. It should not result in charging any objective with a greater investment than the fair capitalized value of the annual benefit of this objective to the consumer. It should not result in charging any objective with a greater investment than would suffice for its development at an alternate single purpose site. Finally, it should not charge any two or more objectives with a greater investment than would suffice for alternate dual purpose or multiple purpose improvement."

The allocation problem which confronted the Authority was unique because it was the first federal agency to encounter a problem of this nature on a large scale and hence, the TVA did not find at hand a generally acceptable solution. In Section IV.1 we shall treat several allocation methods which were studied by the TVA.

## 2. Cooperative games in characteristic function form

In the TVA allocation problem, a fair apportionment of joint project investment cost among several objectives had to be found. As in the first preliminary criterion of Ransmeier (1942), criteria of a fair allocation can be drawn up by taking into account for any subgroup of objectives the given alternative cost of a project designed solely for the objectives in the involved subgroup. So, to any subgroup there corresponds a single numerical index. Hence, there is available a real-valued function on the family of subsets of a finite set. This function is the cornerstone of the theory concerning the cooperative games in characteristic function form.

DEFINITION. Let $n \in \mathbb{N}$.
A _cooperative n-person game in characteristic function form_ is an ordered pair $(N;v)$, where $N$ is a set of $n$ elements and $v\colon 2^N \to \mathbb{R}$ is a real-valued set-function on the set $2^N$ of all subsets of $N$ such that $v(\emptyset) = 0$.

Elements of the set $N$ are called _players_ and the relevant set-function $v$ the _characteristic function_ of the game.
A subset $S$ of the player set $N$ (notation: $S \subset N$) is called a _coalition_ and $v(S)$ the _worth of coalition_ $S$ in the game. The player set $N$ itself is also called the grand coalition, whereas a coalition $S$ is said to be nontrivial if $S \neq N,\emptyset$. The number of players in a coalition $S$ is denoted by $|S|$.
Generally, we shall identify the cooperative game $(N;v)$ with its characteristic function $v$. Further, the class of all cooperative n-person games with player set $N$ is denoted by $G^N$.
In general, we shall suppose that the $n$ players in $N$ are numbered by 1, 2, ••• and n. So, unless stated otherwise, throughout this work we suppose $N = \{1,2,\ldots,n\}$ and we also write $G^n$ instead of $G^N$.

In many cases, the elements of the player set N represent real persons, e.g., landowners and peasants (see Section 3), traders (see 4), creditors (see 6) or voters (see 8), but the player set can also consist of objectives as in the well-known TVA case, airport landings by planes (see 5) or agricultural associations and city water services (see 7). In the next sections, it is further described how the characteristic function can be determined. There the worth of coalition S represents the monetary value of the output produced by cooperation between its members (see 3 and 4), the cost of a project designed solely for the members of S (see 5 and 7) or the abstract power of coalition S in a voting system (see 8).

The practical examples of the next sections are meant to illustrate some parts of the theory developed in this work. In the numerical examples we shall delete the braces and write v(1), v(12), v(137) instead of v({1}), v({1,2}), v({1,3,7}), etc.

## 3. A production economy with landowners and peasants

We consider a production economy in which several landless peasants and one (two) landowner(s) are involved. The particular production model (mainly with one landowner) has already been studied in Shapley and Shubik (1967) and in Chetty, Dasgupta and Raghavan (1976).
For this economic model we suppose that the two landowners are of the same type. Moreover, the peasants have nothing to contribute but their labour and are also supposed to be of the same type. The landowners hire peasants to cultivate their land. If t peasants are hired by one landowner, then the monetary value of the crop of the land cultivated by these t peasants is denoted by $f(t) \in \mathbb{R}$. The function $f: \{0,1,\ldots,m\} \to \mathbb{R}$ is called the <u>production</u> <u>function</u> where m is the total number of peasants ($m \geq 1$). Throughout the following, it is required that the production function f satisfies the next two conditions:

(i) a landowner by himself can not produce anything, i.e., $f(0) = 0$.
(ii) f is a nondecreasing function, i.e.,
$f(t+1) \geq f(t)$ for all $t \in \{0,1,\ldots,m-1\}$.

The two conditions imply that f is a nonnegative function, i.e., $f(t) \geq 0$ for all $t \in \{0,1,\ldots,m\}$.

For the economic situation containing only one landowner, we regard the landowner as player 1 and the peasants as players 2, •••, m+1. Then this situation can be modelled as a cooperative (m+1)-person game (N;v) where its player set $N := \{1,2,...,m+1\}$ and its characteristic function $v: 2^N \to \mathbb{R}$ is given by

$$v(S) := 0 \qquad \text{if } 1 \notin S$$
$$= f(|S|-1) \qquad \text{if } 1 \in S.$$

The worth of any coalition consisting of peasants only equals zero because the peasants do not own any land. Further, the worth of any coalition containing the landowner equals the monetary value of the crop of the land cultivated by the peasants in the involved coalition. Note that $v(\{i\}) = 0$ for all $i \in N$.

For the economic situation containing both landowners, we regard the landowners as players 1, 2, and the peasants as players 3, •••, m+2. Let $M := \{1,2,...,m+2\}$. Then the characteristic function of the corresponding game (M;w) is given by

$$\begin{aligned} w(S) &:= 0 && \text{if } 1 \notin S \text{ and } 2 \notin S \\ &= f(|S|-1) && \text{if } 1 \in S,\ 2 \notin S \\ & && \text{or } 1 \notin S,\ 2 \in S \\ &= \max\left(f(r)+f(t) \mid r\geq 0, t\geq 0, r+t=|S|-2\right) && \text{if } \{1,2\} \subset S. \end{aligned}$$

Notice that, in case the two landowners are members of a coalition, then the peasants in the involved coalition are divided into two disjoint groups in such a way that the largest possible output from the coalition is attained. In addition, $w(\{i\}) = 0$ for all $i \in M$. The two games v and w corresponding to the above economic situations are treated in Example V.2.2.

## 4. An exchange economy with traders of two types

We consider an exchange economy consisting of several traders of two types and of two completely complementary commodities A and B which are usable only in equal quantities. Thus, the set N of traders is divided into two disjoint nonempty subsets P and Q, where traders in P (Q respectively) each initially hold one ($\alpha \in \mathbb{R}$ where $\alpha \geq 0$) unit(s) of the commodity A (B). Furthermore, it is supposed that the output produced with one unit of both commodities can be sold at a net profit of one unit of money. The net profit function v which describes for any subset of traders the largest possible monetary value of the output of the goods possessed by the involved subset of traders, is given by

$$v(S) := \min(|S \cap P|, \alpha|S \cap Q|) \qquad \text{for all } S \subset N. \quad (1.1)$$

This economic situation can be modelled as a cooperative game $(N;v)$ where its player set $N = P \cup Q$ consists of the traders and its characteristic function v is precisely the net profit function. Note that $v(\{i\}) = 0$ for all $i \in N$.

This type of a game where $\alpha = 1$ is also known as a glove game (cf. Rosenmüller, 1971, page 13). The traders in P (Q respectively) possess a left (right) hand glove and so, the number of pairs of gloves that can be formed within a subset S of traders is then given by the expression v(S) of (1.1).

The case $\alpha = 0.5$, $|P| = 2$ and $|Q| = 3$ of this type of a game can also be interpreted as follows (Maschler, 1976): "Each of two manufacturers owns two machines that can be operated only by skilled workers. There are exactly three available skilled workers, each willing to work at most 8 hr a day. When a worker operates the machine during 8 hr he produces an item that can be sold at a net profit of half a unit of money."

The game v arising from the exchange economy consisting of three traders where $0.5 \le \alpha \le 1$, $|P| = 1$ and $|Q| = 2$, is treated in Example II.4.4.

## 5. The airport game

As an application of game theoretic analysis to the cost allocation problem, Littlechild and Owen (1973) considered the problem of setting airport landing charges for different types of aircraft.

In general, there are two kinds of expenses at an airport:

(i) variable operating costs due to the landings by planes of different types and

(ii) a fixed capital cost (e.g., terminal construction and runway construction).

The variable operating costs are directly traceable to the planes using the airport and hence, these costs are charged on a per-landing basis. Consequently, the cost allocation problem becomes the problem of how to allocate the capital cost to the planes. The fixed capital cost of providing an airport runway essentially depends upon the largest plane because a new larger plane which is to land at the airport, needs an enlarged runway.

We divide the planes into m types ($m \geq 1$). Let $N_j$ be the set of landings by planes of type j ($j = 1, \cdots, m$) and $N := \bigcup_{j=1}^{m} N_j$ the set of all landings at the airport. Let $C_j$ be the cost of a runway adequate for planes of type j. Without loss of generality, these types may be ordered so that $0 = C_0 < C_1 < C_2 < \cdots < C_m$.

Let $S \subset N$, $S \neq \emptyset$. Then the cost $c(S)$ of a runway adequate to receive all landings in S is given by

$$c(S) := \max\{C_j \mid 1 \leq j \leq m,\ S \cap N_j \neq \emptyset\}$$

while $c(\emptyset) := 0$. Both the airport cost allocation problem and a numerical example concerning the Birmingham airport (Birmingham, U.K.) during the year 1968-1969, are treated in Section IV.4.

## 6. The bankruptcy game

The Babylonian Talmud is a 2,000-year old document that forms the basis for Jewish religious, criminal as well as civil law. It includes two kinds of teachings:

(i) the Mishna which are short statements of the law and

(ii) the Gemara which are commentary on the Mishna by the rabbis of that time.

Around the year 1140 A.D., rabbi Ibn Ezra gave the next problem.

"Jacob died and each of his four sons Reuben, Simeon, Levi and Judah respectively produced a deed that Jacob willed to him his entire estate, half, one third, one quarter of his estate on his death. All deeds bear the same date and the total estate is 120 units."

A _general bankruptcy problem_ is defined as an ordered pair $(E;d)$, where $E \in \mathbb{R}$ and $d = (d_1, d_2, \ldots, d_n) \in \mathbb{R}^n$ such that

$$d_i \geq 0 \text{ for all } 1 \leq i \leq n \qquad \text{and} \qquad 0 \leq E \leq \sum_{j=1}^{n} d_j .$$

The reader may think of a person who dies, leaving debts $d_1, d_2, \dots, d_n$, totalling at least as much as his estate E. The problem is that the debts are mutually inconsistent in that the estate is insufficient to meet all of the debts.

The game theoretic approach to the bankruptcy problem is started in O'Neill (1982) who defined the corresponding <u>bankruptcy game</u> $(N;v_{E;d})$ by $N := \{1,2,\dots,n\}$ and

$$v_{E;d}(S) := \max\Big(0,\ E - \sum_{j\in N-S} d_j\Big) \quad \text{for all } S \subset N. \quad (1.2)$$

So, the player set N consists of the n creditors (or heirs) and the worth of coalition S equals either zero or what is left of the estate after each member i of the complementary coalition N-S is paid his associated claim $d_i$.

Particularly, it follows immediately from the formula (1.2) that the characteristic function of the bankruptcy game possesses the next four properties:

(i) $v_{E;d}(N) = E;$ and for all $i \in N$

(ii) $v_{E;d}(N-\{i\}) = \max(0,\ E - d_i);$

(iii) $v_{E;d}(\{i\}) = \max\Big(0,\ E - \sum_{j\in N-\{i\}} d_j\Big) \le d_i;$

(iv) $d_i \ge E$ implies $v_{E;d}(S) = 0$ for all $S \subset N-\{i\}$.

The above bankruptcy problem with the estate $E = 120$ and the four claims $d_1 = 30$, $d_2 = 40$, $d_3 = 60$, $d_4 = 120$, can be modelled as the 4-person bankruptcy game $(N;v)$ given by

$$\begin{array}{lll} v(14) = 20, & v(24) = 30, & v(34) = 50, \\ v(124) = 60, & v(134) = 80, & v(234) = 90, \\ v(N) = 120, & v(S) = 0 & \text{for all other } S \subset N. \end{array}$$

The game theoretic analysis of the bankruptcy problem is treated in Chapter VI.

## 7. Cooperative water resource development in Japan

In Japan, the set N of agents who exploit the water resources in a given area can be partitioned into two distinct subsets: the subset P consisting of agricultural associations who view the existing water supplies as being (more than) adequate for their own irrigation needs, and the subset Q consisting of city water service authorities whose current or future water needs are not met by existing sources. Each city might acquire the quantity of additional water it needs in two ways:

(i) construct a dam with or without the cooperation of other cities or

(ii) arrange with the agricultural associations for the direct diversion of water from them to the city.

A city may also opt for a combination of (i) and (ii) to meet its needs. Thus, any further exploitation of the water resources to meet the needs of the cities in the area can be achieved in case the agents form a coalition to carry out the necessary construction projects. Obviously, the crucial motive of the agents to cooperate would be based on the fact that the cost involved in so doing would be less than the sum of the costs of each city alone meeting its own additional needs.

This cooperative development of a water resource can be modelled as a cooperative game with the set N of agents as player set. In order to define its characteristic (cost) function c, let S be any nonempty coalition of agricultural associations and/or city water service authorities. Then c(S) represents the minimum cost of meeting the additional water needs of the cities in S, on the understanding that there is no cooperation from those agents outside S.

Thus, in case $S \cap Q = \emptyset$, then $c(S) = 0$ as the agricultural associations have no additional water needs. Moreover, in case $S \cap P = \emptyset$, then c(S) must be the cost of constructing a dam just capable of meeting the additional water needs of the cities in S. For more details about the determination of the numerical cost c(S) for each possible coalition S, we refer to Suzuki and Nakayama (1976). Their numerical example with two agricultural associations and three city water service authorities is treated in Section IV.3. There several cost allocations for the cost game of the example are compared.

## 8. Simple games

Game theory can be used to describe the abstract power of a voter in voting systems because of the invention of simple games in Von Neumann and Morgenstern (1944). These simple games are completely characterized by the fact that the coalitions in the game can be divided into two types:

winning (powerful) and losing (powerless) coalitions. An n-person game v is said to be <u>simple</u> if

$$v(S) \in \{0,1\} \quad \text{for all } S \subset N, \quad v(N) = 1, \quad \text{and}$$
$$v(S) \leq v(T) \quad \text{for all } S \subset T \subset N. \tag{1.3}$$

The condition (1.3) is known as the <u>monotonicity</u> condition for the n-person game v. The class of simple n-person games is denoted by $S^n$. The players whose absence gives rise only to losing coalitions are of the utmost importance and are called <u>veto players</u>. Thus, the set $J^v$ of all veto players in a simple n-person game v is defined by

$$J^v := \{i \in N \mid v(N-\{i\}) = 0\}.$$

An example of a simple game is as follows. Let $m \in \mathbb{N}$ and $n \in \mathbb{N}$ be such that $1 \leq m < n$. Consider a council consisting of n members including one chairman. In order to pass a bill, at least m votes are needed including the vote of the chairman. The simple n-person game v which corresponds to this voting situation is given by

$$\begin{aligned} v(S) &= 1 \quad \text{if } 1 \in S \text{ and } |S| \geq m \\ &= 0 \quad \text{otherwise} \end{aligned} \tag{1.4}$$

where player 1 represents the chairman. Note that $J^v = \{1\}$. The game of (1.4) where $m = 1$ is called the unanimity game with respect to player 1. In general, the unanimity game with respect to a nonempty coalition represents the voting system in which the members of the involved coalition play a prominent part because their votes are needed in order to pass a bill.
That is, for any $T \subset N$, $T \neq \emptyset$, the <u>unanimity game</u> $u_T \in G^n$ is defined by

$$\begin{aligned} u_T(S) &:= 1 \quad \text{if } S \supset T \\ &= 0 \quad \text{otherwise.} \end{aligned} \tag{1.5}$$

Thus, a coalition is powerful (powerless respectively) in the simple unanimity game $u_T$ if it does (not) include all members of T and its associated worth is put to one (zero). From this, it follows that the members of T are precisely the veto players in the unanimity game $u_T$, i.e., $J^{u_T} = T$. Simple games are also treated in Section III.7.

## 9. Notions

In this final section we treat some notions as well as some notation which will be used throughout this work. Here a set-function $v: 2^N \to \mathbb{R}$ with $v(\emptyset) = 0$ or equivalently, the n-person game v is called

| | | |
|---|---|---|
| zero-normalized | if $v(\{i\}) = 0$ | for all $i \in N$ |
| monotonic | if $v(S) \le v(T)$ | for all $S \subset T \subset N$ |
| zero-monotonic | if $v(S) + \sum_{j \in T-S} v(\{j\}) \le v(T)$ | for all $S \subset T \subset N$ |
| additive | if $v(S) = \sum_{j \in S} v(\{j\})$ | for all $S \subset N$ |
| superadditive | if $v(S) + v(T) \le v(S \cup T)$ | for all $S,T \subset N$ with $S \cap T = \emptyset$. |

The superadditivity notion states that it is advantageous (in terms of savings) for disjoint coalitions to form their union, while the monotonicity notion expresses that the worth does not decrease whenever the coalition is enlarged. Clearly, superadditivity implies zero-monotonicity.
Let $v \in G^n$, $w \in G^n$ and $\alpha \in \mathbb{R}$. Then the n-person games $v + w$ and $\alpha v$ are defined by

$$(v + w)(S) := v(S) + w(S) \qquad \text{for all } S \subset N,$$
$$(\alpha v)(S) := \alpha v(S) \qquad \text{for all } S \subset N.$$

With respect to this addition and multiplication, the class $G^n$ of n-person games is a $(2^n-1)$-dimensional linear space. A basis of $G^n$ is given by the set $\{u_T \in G^n | \ T \subset N, T \neq \emptyset\}$ of all unanimity n-person games with player set N. The linear independence of this set is shown in the proof of Theorem V.1.2. The game v is said to be equal to the game w (notation: $v = w$) whenever $v(S) = w(S)$ for all $S \subset N$.

Let $v \in G^n$, $\alpha \in (0,\infty)$, $d \in \mathbb{R}^n$ and $\theta: N \to N$ a permutation. The n-person games $\alpha v + d$ and $\theta v$ are defined by

$$(\alpha v + d)(S) := \alpha v(S) + \sum_{j \in S} d_j \qquad \text{for all } S \subset N,$$

$$(\theta v)(\theta S) := v(S) \qquad \text{for all } S \subset N.$$

The n-person game $\alpha v + d$ where $\alpha = 1$ and $d_i = -v(\{i\})$ for all $i \in N$, is called the zero-normalization of the game v because of $(\alpha v + d)(\{i\}) = 0$ for all $i \in N$. The n-person

game $\theta v$ represents the game v after the renumbering of the players by means of the ordering $\theta$ of the player set N.
Let $i \in N$ and $j \in N$ be such that $i \neq j$. Then the set of all coalitions containing player i but not player j, is denoted by $\Gamma_{ij}$. Thus,

$$\Gamma_{ij} := \{S \subset N \mid i \in S,\ j \notin S\}.$$

Let $v \in G^n$, $x \in \mathbb{R}^n$ and $S \subset N$. We usually write $x(S)$ instead of $\sum_{j \in S} x_j$ where $x(\emptyset) := 0$. The <u>excess of the coalition</u> S <u>with respect to the vector</u> x in the game v is defined to be

$$e^v(S,x) := v(S) - \sum_{j \in S} x_j = v(S) - x(S). \tag{1.6}$$

A nonnegative (nonpositive respectively) excess of S at x in the game v represents the gain (loss) to the coalition S if its members withdraw from the payoff vector x in order to form their own coalition. We often write $e(S,x)$ instead of $e^v(S,x)$. As we shall see in the next chapter, the idea of excess forms the basis of several solution concepts for cooperative games in characteristic function form.

# CHAPTER II

# SOLUTION CONCEPTS FOR COOPERATIVE GAMES AND RELATED SUBJECTS

## 1. Notions

With the characteristic function v at hand and supposing that some type of understanding is arrived at by the players, they have to divide the total savings v(N) of their grand coalition. A distribution of the amount v(N) among the players will be represented by a real-valued function x on the player set N satisfying the <u>efficiency principle</u> $\sum_{j \in N} x(j) = v(N)$. Here x(i) which is also denoted by $x_i$, represents the payoff to player i according to the involved payoff function x. Because we generally suppose that the player set $N = \{1,2,\ldots,n\}$, we usually identify a real-valued function $x \in \mathbb{R}^N$ on N with the n-tuple $x = (x_1,x_2,\ldots,x_n) \in \mathbb{R}^n$ of real numbers. The vectors $x \in \mathbb{R}^n$ which satisfy the efficiency principle $x(N) = v(N)$ are called <u>efficient payoff vectors</u> or <u>pre-imputations</u> for the n-person game v. The nonempty set of all pre-imputations is denoted by $I^*(v)$, i.e.,

$$I^*(v) := \{x \in \mathbb{R}^n \mid x(N) = v(N)\}. \tag{2.1}$$

Since the introduction of the notion of a cooperative n-person game in characteristic function form, many solution concepts for these games have been proposed. The solution concepts prescribe somehow a specific subset of the pre-imputation set. Formally, a <u>solution concept</u> on a nonempty collection G of games is a (multi)function $\psi$ on G which associates with any game $v \in G$ a subset $\psi(v)$ of its pre-imputation set $I^*(v)$. Note that the associated set $\psi(v)$ with a game $v \in G$ is allowed to be empty.

Most of the proposed solution concepts meet the <u>individual rationality principle</u> which requires that the payoff to any player i by a payoff vector x is at least the amount what player i can attain for himself in the game v, i.e., $x_i \geq v(\{i\})$ for all $i \in N$. The pre-imputations which also satisfy the individual rationality principle are called <u>imputations</u> for the n-person game v. The set of all imputations is denoted by I(v), i.e.,

$$I(v) := \{x \in \mathbb{R}^n \mid x(N) = v(N) \text{ and } x_i \geq v(\{i\}) \text{ for all } i \in N\}. \tag{2.2}$$

In this chapter we proceed with a general treatment of the stable sets (see Section 3), the balancedness (see 8) and the following solution concepts:
- the core and the strong $\epsilon$-cores (see 4)
- the bargaining set (see 5)
- the kernel and the prekernel (see 6); and further
- the Shapley value (see 2)
- the nucleolus (see 7)
- the $\tau$-value (see Chapter III)

where the last three solution concepts select a unique pre-imputation. In the definitions of solution concepts involving the imputation set I(v) of a game v, we tacitly suppose $I(v) \neq \emptyset$. It is left to the reader to verify that

$$I(v) \neq \emptyset \quad \text{if and only if} \quad v(N) \geq \sum_{j \in N} v(\{j\}).$$

In view of the "one-point" solution concepts, we list several desirable properties for values on any nonempty collection G of games. Here a <u>value</u> on G is a function $\psi$ on G such that $\psi(v) \in I^*(v)$ for all $v \in G$. The i-th coordinate $\psi_i(v)$ of the vector $\psi(v)$ represents the value of player i in the game $v \in G$.

(i) <u>Individual rationality</u>.
For all $(N;v) \in G$ and all $i \in N$: $\psi_i(v) \geq v(\{i\})$.

(ii) <u>Symmetry</u>.
For all $(N;v) \in G$ and any permutation $\theta: N \to N$ with $(N;\theta v) \in G$: $\psi_{\theta(i)}(\theta v) = \psi_i(v)$ for all $i \in N$.

(iii) <u>Dummy player property</u>.
For all $(N;v) \in G$ and any dummy player $i \in N$ in the game $(N;v)$: $\psi_i(v) = v(\{i\})$.

Here player i is called a <u>dummy</u> in the game (N;v) if

$$v(S \cup \{i\}) - v(S) = v(\{i\}) \quad \text{for all } S \subset N-\{i\}.$$

(iv) Relative invariance under S-equivalence.
For all $(N;v) \in G$, $\alpha \in (0,\infty)$ and $d \in \mathbb{R}^n$ with $(N;\alpha v + d) \in G$: $\psi(\alpha v + d) = \alpha\psi(v) + d$.

The symmetry property expresses that a renumbering of the players does not affect the values of the players. A dummy is a player whose marginal contribution to any coalition is always equal to the worth of his own coalition and hence, according to the dummy player property, his value equals his own worth. The relative invariance under S-equivalence requires that the value behaves in a natural way with respect to changes in scale which are comparable with positive affine transformations. Finally, we formulate the additivity condition for a value which requires that the sum of the values of two separate games does not differ from the value of the natural sum game of the two involved games.

(v) Additivity.
For all $(N;v) \in G$ and $(N;w) \in G$ with $(N;v + w) \in G$: $\psi(v + w) = \psi(v) + \psi(w)$.

## 2. The Shapley value

The Shapley value is a specific "one-point" solution concept which was introduced and characterized in Shapley (1953) with the aid of several properties, including the additivity. The familiar formula for the Shapley value $\phi(v) \in \mathbb{R}^n$ of a game $v \in G^n$ is as follows (Shapley, 1953): for all $i \in N$

$$\phi_i(v) := \sum_{S\subset N-\{i\}} \gamma_n(S)\left[v(S \cup \{i\}) - v(S)\right] \tag{2.3}$$

where

$$\gamma_n(T) := (n!)^{-1}|T|!(n-|T|-1)! \tag{2.4}$$

for all $T \subset N$, $T \neq N$. In view of the formula (2.4), we have

$$\gamma_n(T) = n^{-1}\binom{n-1}{|T|}^{-1} \quad \text{and} \quad \sum_{S\subset N-\{i\}} \gamma_n(S) = 1 \text{ for all } i \in N.$$

This implies that $\{\gamma_n(S) \mid S \subset N-\{i\}\}$ can be regarded as a probability distribution over the collection of subsets of N not containing player i. Notice that this probability distribution arises from the belief that the coalition, to which player i joins, is equally likely to be of any size t, $0 \le t \le n-1$, and that all such coalitions of the same size t are equally likely. If, for each $T \subset N-\{i\}$, $\gamma_n(T)$ is seen as the probability that player i joins the coalition T and the marginal contribution $v(T \cup \{i\}) - v(T)$ is paid to player i for joining the coalition T, then the Shapley value $\phi_i(v)$ for player i, as given by the formula (2.3), is simply the expected payoff to player i in the game v.

First of all, we show that the Shapley value of a game is indeed an efficient payoff vector for the game by introducing an alternative formula for the Shapley value, in which the marginal contributions of a player to the feasible coalitions are replaced by the power of those coalitions with respect to their complementary coalitions.

PROPOSITION 2.1. For all $v \in G^n$ and all $i \in N$

$$\phi_i(v) = \sum_{S \subset N-\{i\}} \gamma_n(S)\big(v(N-S) - v(S)\big). \tag{2.5}$$

PROOF. Let $v \in G^n$ and $i \in N$. Consider $S \subset N-\{i\}$ and put $S^c := N-S$. Then we also have $S^c-\{i\} \subset N-\{i\}$ and by (2.4), $\gamma_n(S^c-\{i\}) = \gamma_n(S)$. Now it follows that

$$\gamma_n(S)\big(v(S\cup\{i\}) - v(S)\big) + \gamma_n(S^c-\{i\})\big(v(S^c) - v(S^c-\{i\})\big) =$$
$$\gamma_n(S)\big(v(S^c) - v(S)\big) + \gamma_n(S^c-\{i\})\big(v(S\cup\{i\}) - v(S^c-\{i\})\big).$$

Therefore, the formula (2.5) is a direct consequence of the formula (2.3). □

THEOREM 2.2. $\sum_{j=1}^{n} \phi_j(v) = v(N)$ for all $v \in G^n$.

PROOF. Let $v \in G^n$. For any player $i \in N$ we define the set-function $w_{N-\{i\}}: 2^N \to \{0,1\}$ by

$$w_{N-\{i\}}(S) := 1 \quad \text{if } S \subset N-\{i\}$$
$$= 0 \quad \text{otherwise.}$$

We write $S \subsetneq N$ instead of $S \subset N$, $S \neq N$. For any $S \subsetneq N$, put $h(S) := \gamma_n(S)\big(v(N-S) - v(S)\big)$. It follows from (2.5) that

$$\sum_{j=1}^{n} \phi_j(v) = \sum_{j\in N} \sum_{S\subset N-\{j\}} h(S) = \sum_{j\in N} \sum_{S\subseteq N} w_{N-\{j\}}(S)\, h(S)$$

$$= \sum_{S\subseteq N} \sum_{j\in N} w_{N-\{j\}}(S)\, h(S) = \sum_{S\subseteq N} \sum_{j\in N-S} h(S)$$

$$= \sum_{S\subseteq N} (n - |S|)h(S) = \sum_{S\subseteq N} \binom{n}{|S|}^{-1}\big(v(N-S) - v(S)\big).$$

Note that, in case $S \subset N$, $S \neq \emptyset$, then also $N-S \subseteq N$, while $\binom{n}{|S|} = \binom{n}{|N-S|}$. We conclude that

$$\sum_{j=1}^{n} \phi_j(v) = \binom{n}{0}^{-1}\big(v(N) - v(\emptyset)\big) = v(N). \qquad \square$$

By the above theorem, the Shapley value possesses the efficiency property, while both the dummy player property and the additivity property for the Shapley value follow immediately from the formula (2.3). It is left to the reader to verify by straightforward calculations that the Shapley value also possesses the symmetry property. As a matter of fact, the Shapley value is completely axiomatized by these four properties which result is due to Shapley (1953).

THEOREM 2.3. The Shapley value $\phi\colon G^n \to \mathbb{R}^n$ is the unique value on $G^n$ with the following three properties: symmetry, dummy player property and additivity.

PROOF. It remains to prove the uniqueness part. Suppose that $\psi\colon G^n \to \mathbb{R}^n$ is a value with the three mentioned properties. For any $T \subset N$, $T \neq \emptyset$, consider the unanimity n-person game $u_T$ as given by (1.5). Let $T \subset N$, $T \neq \emptyset$, and $\alpha \in \mathbb{R}$. Any $i \in N-T$ is a dummy player in the n-person game $\alpha u_T$ and hence, by the dummy player property for $\psi$, we have

$$\psi_i(\alpha u_T) = \alpha u_T(\{i\}) = 0 \qquad \text{whenever } i \in N-T.$$

Let $j \in T$ and $k \in T$. There exists a permutation $\theta\colon N \to N$ satisfying $\theta(j) = k$ and $\theta(T) = T$. Then $\theta(\alpha u_T) = \alpha u_T$ and hence, by the symmetry property for $\psi$, we obtain

$$\psi_k(\alpha u_T) = \psi_{\theta(j)}\big(\theta(\alpha u_T)\big) = \psi_j(\alpha u_T) \qquad \text{whenever } j,\ k \in T.$$

Further, $\sum_{r\in N} \psi_r(\alpha u_T) = \alpha u_T(N) = \alpha$ because of the efficiency property for $\psi$. Now it follows that

$$\psi_i(\alpha u_T) = 0 \qquad \text{if } i \in N-T$$
$$= |T|^{-1}\alpha \qquad \text{if } i \in T.$$

We conclude that $\psi(\alpha u_T)$ is unique for any $T \subset N$, $T \neq \emptyset$, and $\alpha \in \mathbb{R}$. Recall that the set $\{u_T \in G^n | \ T \subset N, T \neq \emptyset\}$ of all unanimity games forms a basis of the linear space $G^n$. Together with the additivity property for $\psi$, this implies that $\psi(v)$ is unique for any $v \in G^n$. Thus, if $\psi$ exists, then $\psi$ is unique. □

By the formula (2.3), it is obvious that the Shapley value is also relatively invariant under S-equivalence. However, the Shapley value on $G^n$ fails to be individually rational. Nevertheless, the Shapley value of a superadditive game is always an imputation.
Other axiomatic characterizations of the Shapley value on $G^n$ which include the additivity property can be obtained by replacing the classical symmetry and/or dummy player properties by other properties (cf. Aumann and Shapley, 1974, page 299; Driessen, 1985a). Axiomatizations of the Shapley value which do not include the additivity property are given in Young (1985a) who characterized the Shapley value on $G^n$ by means of a certain monotonicity property, and in Sobolev (1973) who characterized the Shapley value on the class of all games by means of a so-called reduced game property.
The restriction of the Shapley value on $G^n$ to the class $S^n$ of simple n-person games is called the Shapley-Shubik power index. This power index was introduced in Shapley and Shubik (1954) and it can not be characterized by means of the additivity property because the class $S^n$ is not closed under addition. Axiomatic characterizations and treatments of the Shapley-Shubik power index and other power indices which represent quantitative measures for evaluating the power of a voter, are given in various papers, e.g., Banzhaf (1965); Bolger (1980, 1982); Coleman (1971); Curiel (1987); Deegan and Packel (1978, 1983); Dubey (1975a,b); Dubey and Shapley (1979); Lucas (1983); Owen (1975a, 1978a,b); Roth (1977); Straffin (1983) and Weber (1978).
In Section V.4 we pay attention to another formula for the Shapley value in terms of the orderings of the player set.

## 3. The stable sets

The idea of the stable sets was first introduced in Von Neumann and Morgenstern (1944). The stable sets (or equivalently, the Von Neumann - Morgenstern solutions) are described in terms of a relation between imputations called domination.

DEFINITION 3.1. Let $v \in G^n$, $x \in I(v)$ as well as $y \in I(v)$. We say x <u>dominates</u> y (notation: x dom y) if there exists a nonempty coalition S such that

$$x_i > y_i \text{ for all } i \in S \quad \text{and} \quad \sum_{j \in S} x_j \le v(S). \quad (2.6)$$

The first condition requires that all players in S prefer the imputation x to y, while the second condition requires that their savings by cooperation in the game is at least their total payoff according to the imputation x. Notice that the condition (2.6) excludes the domination through the grand coalition and the one-person coalitions.
Informally, a stable set V satisfies the two conditions of internal stability (no imputation in V dominates another) and external stability (any imputation outside the set V is dominated by some imputation in V). Formally, the two stability conditions are described as follows.

DEFINITION 3.2. Let $v \in G^n$. A set $V \subset I(v)$ is said to be a <u>stable set</u> for the game v if it satisfies the next two conditions:
(i) If $x \in V$ and $y \in V$, then not x dom y.
(ii) If $x \in I(v) - V$, then there exists $y \in V$ such that y dom x.

Another formulation of the stability of a set $V \subset I(v)$ is as follows. Denote by dom V the set of all imputations which are dominated by imputations in V, i.e.,

$$\text{dom } V := \{x \in I(v) \mid \text{there exists } y \in V \text{ such that } y \text{ dom } x\}.$$

Then the set V is stable if and only if the imputation set can be partitioned into the two subsets V and dom V. That is, $V \subset I(v)$ is a stable set if and only if

$$V \cap \text{dom } V = \emptyset \quad \text{and} \quad V \cup \text{dom } V = I(v).$$

The straightforward proof of the above equivalence is left to the reader.

Lucas (1968, 1969) described a ten-person game without a stable set and consequently, existence of stable sets is not guaranteed. Nevertheless, many games possess immense collections of stable sets, but there are also games with a unique stable set. In Section V.5 we consider a certain condition which guarantees the uniqueness of stable sets. In general, the determination of the stable sets for a game is very laborious. A generalization of a stable set, called a subsolution, is given in Roth (1976). By the definitions, every stable set is also a subsolution.

## 4. The core and the strong $\epsilon$-cores

The first preliminary criterion of a satisfactory cost allocation as stated in Ransmeier (1942) and mentioned in Section I.1, foreshadowed the idea of the core of a game which was first introduced and named in game theory in Gillies (1953) as an adjunct to studies of the stable sets. Due to the fact that the core of a game may be empty, Shapley and Shubik (1963, 1966) introduced the notion of a strong $\epsilon$-core as a generalization of the core.

DEFINITION 4.1. For any $v \in G^n$ and $\epsilon \in \mathbb{R}$, the <u>strong $\epsilon$-core</u> $C_\epsilon(v)$ of the game v is given by

$$C_\epsilon(v) := \{x \in \mathbb{R}^n \mid x(N) = v(N) \text{ and } x(S) \geq v(S) - \epsilon \text{ for all } S \neq N,\emptyset\}. \tag{2.7}$$

In particular, the <u>core</u> $C(v)$ of a game $v \in G^n$ is given by

$$C(v) := \{x \in \mathbb{R}^n \mid x(N) = v(N) \text{ and } x(S) \geq v(S) \text{ for all } S \subset N\}. \tag{2.8}$$

Clearly, $C_0(v) = C(v)$ and $C(v) \subset I(v)$ for all games $v \in G^n$.

For any nonnegative (nonpositive respectively) real number $\epsilon$, the strong $\epsilon$-core of a game can be interpreted as the set of all pre-imputations that can not be improved upon by any coalition if one imposes a cost of $\epsilon$ (bonus of $-\epsilon$) in all cases where a nontrivial coalition is formed. In view of (1.6), we can also state that the strong $\epsilon$-core of a game consists of pre-imputations that give rise only to excesses not greater than $\epsilon$ for all nontrivial coalitions.

The relationship of the core to the stable sets will be derived from the next theorem. According to part (i) of the theorem, the core of any game is included in the set of all undominated imputations for the game (and therefore, the core is always internally stable). In general, this inclusion is strict, but part (ii) of the theorem states that for superadditive games the inclusion is an equality. Part (ii) is due to Shapley and Shubik (1969).

THEOREM 4.2. Let $v \in G^n$.

(i) If $x \in C(v)$, then there exists no $y \in I(v)$ such that $y$ dom $x$.

(ii) If the game $v$ is superadditive, then
$C(v) = \{x \in I(v) \mid \text{there is no } y \in I(v) \text{ with } y \text{ dom } x\}$.

PROOF. (i) Let $x \in C(v)$. Assume that there exists $y \in I(v)$ such that $y$ dom $x$. By (2.6), there exists $S \subset N$, $S \neq \emptyset$, with $x(S) < y(S) \leq v(S)$. Thus, $x(S) < v(S)$ which strict inequality is in contradiction with $x \in C(v)$ by (2.8). So, the statement in part (i) holds.
(ii) Let the game $v$ be superadditive. Let $x \in I(v) - C(v)$. We show that there exists $y \in I(v)$ such that $y$ dom $x$. (2.9)
Since $x \in I(v) - C(v)$, there exists $S \subset N$, $S \neq N, \emptyset$, with $x(S) < v(S)$. Put

$$\alpha := v(S) - x(S) \quad \text{and} \quad \beta := v(N) - v(S) - \sum_{j \in N-S} v(\{j\}).$$

Define the vector $y \in \mathbb{R}^n$ by

$$y_i := x_i + |S|^{-1}\alpha \qquad \text{if } i \in S$$
$$\quad = v(\{i\}) + |N-S|^{-1}\beta \qquad \text{if } i \in N-S.$$

Then $y(N) = v(N)$, $y(S) = v(S)$ and $y_i > x_i$ for all $i \in S$ because of $\alpha > 0$. Further, $\beta \geq 0$ by the superadditivity of the game $v$. Now it follows from $x \in I(v)$ and $\beta \geq 0$ that $y_i \geq v(\{i\})$ for all $i \in N$. In view of (2.2) and (2.6), we conclude that $y \in I(v)$ such that $y$ dom $x$. The statement in part (ii) is a direct consequence of (2.9) and part (i). □

COROLLARY 4.3. Let $V$ be a stable set for a game $v \in G^n$.

(i) Then $C(v) \subset V$.

(ii) If the core $C(v)$ is a stable set, then $V = C(v)$.

PROOF. (i) Assume that the inclusion $C(v) \subset V$ does not hold. Then there exists $x \in C(v) - V$. Because $V$ is a stable set, there exists $y \in V$ such that $y$ dom $x$. However, $y$ dom $x$ is in contradiction with $x \in C(v)$ by Theorem 4.2(i).

(ii) Let $C(v)$ be a stable set. By part (i), we always have $C(v) \subset V$. Assume $C(v) \neq V$. Then there exists $x \in V - C(v)$. Since $C(v)$ is a stable set, there exists $y \in C(v)$ such that $y$ dom $x$. We obtain $y$ dom $x$ with $x \in V$ and $y \in C(v) \subset V$. However, this is in contradiction with the (internal) stability of $V$. Hence, $C(v) = V$. □

By the above corollary, every stable set contains the core and in addition, if the core itself happens to be stable, then it is the unique stable set. We remark that the ten-person game, described in Lucas (1968, 1969) in order to show that a game may have no stable sets, has a nonempty core. Lucas and Rabie (1982) demonstrated that there are games of fourteen or more players for which no stable sets exist and for which the core is empty.

Let $v$ be an $n$-person game where $n \geq 2$. Then it is clear that $C_\epsilon(v) \neq \emptyset$ if $\epsilon$ is large enough, while $C_\epsilon(v) = \emptyset$ if $\epsilon$ is small enough. Also, $C_\delta(v) \subset C_\epsilon(v)$ if $\delta < \epsilon$, with strict inclusion if $C_\epsilon(v) \neq \emptyset$. The intersection of all non-empty strong $\epsilon$-cores is called the <u>least-core</u> of the game $v$ and was formally treated for the first time in Maschler, Peleg and Shapley (1979). The least-core $LC(v)$ of the game $v$ can be regarded as revealing the latent position of the core whenever the core of the game $v$ is empty.
We conclude this section with two remarks and an example. It follows immediately from Definition 4.1 that the various strong $\epsilon$-cores, including the core, of an $n$-person game are compact convex subsets of $\mathbb{R}^n$. Consequently, by a well-known theorem of Krein – Milman, any nonempty strong $\epsilon$-core is the convex hull of the set of its extreme points.
Secondly, we remark that another generalization of the core, called the weak $\epsilon$-core, was also presented in Shapley and Shubik (1963, 1969) by considering the expression $v(S) - |S|\epsilon$ instead of $v(S) - \epsilon$ for all $S \neq N,\emptyset$.

EXAMPLE 4.4. Consider the 3-person game $v$ of (1.1) where $P = \{1\}$, $Q = \{2,3\}$ and $0.5 \leq \alpha \leq 1$. By (1.1) and (2.8), we get

$$v(N) = 1, \quad v(12) = v(13) = \alpha, \quad v(S) = 0 \text{ otherwise}, \qquad (2.10)$$

$$C(v) = \{x \in \mathbb{R}^3_+ \mid x_1 + x_2 + x_3 = 1,\ x_2 \leq 1-\alpha,\ x_3 \leq 1-\alpha\}$$
$$= \text{conv}\,\{(1,0,0),\ (\alpha,1-\alpha,0),\ (\alpha,0,1-\alpha),\ (2\alpha-1,1-\alpha,1-\alpha)\}.$$

The core of the game $v$ is drawn in Figure 2.1. Notice that

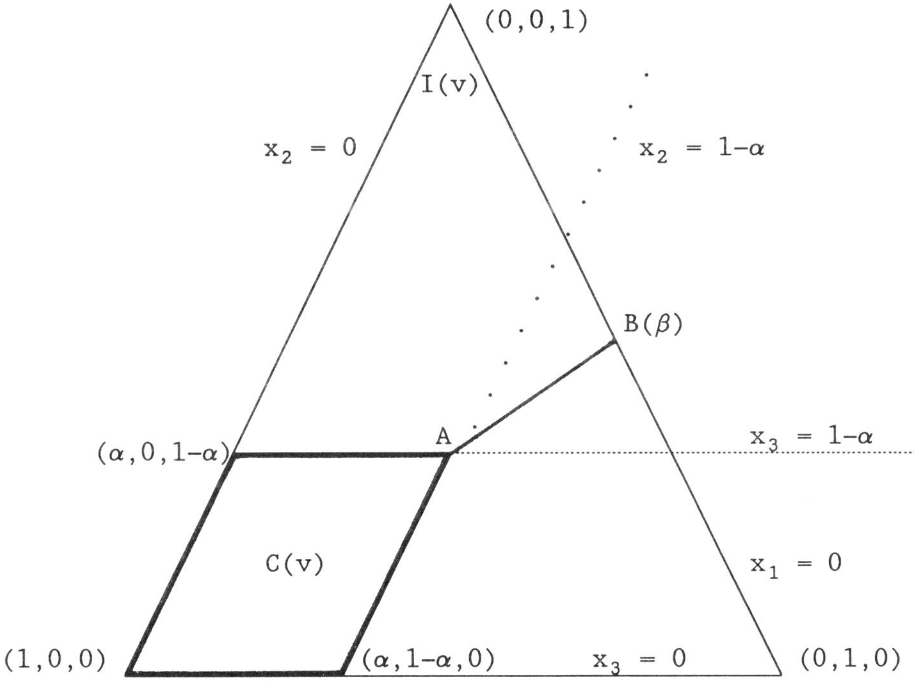


FIGURE 2.1. The indicated quadrilateral represents the core C(v) of the three-person game v of Example 4.4. For any $1-\alpha \le \beta \le \alpha$, the union of the core C(v) and the straight line segment with end points $A = (2\alpha-1, 1-\alpha, 1-\alpha)$ and $B(\beta) = (0, \beta, 1-\beta)$ is a stable set for the game v.

the core is a quadrilateral inside the imputation set, but it degenerates into a singleton whenever $\alpha = 1$. It is left to the reader to verify that

$$C_\epsilon(v) \neq \emptyset \text{ iff } \epsilon \ge \frac{1}{2}(\alpha-1), \quad \text{while} \quad LC(v) = \{\frac{1}{2}(2\alpha, 1-\alpha, 1-\alpha)\}.$$

In order to investigate the stability, we note that domination is only possible through the two-person coalitions {1,2} and {1,3}. By (2.6), we have that for all x, $y \in I(v)$

$$x \text{ dominates } y \text{ through } \{1,2\} \text{ iff } x_1 > y_1,\ x_2 > y_2,\ x_3 \ge 1-\alpha,$$

$$x \text{ dominates } y \text{ through } \{1,3\} \text{ iff } x_1 > y_1,\ x_3 > y_3,\ x_2 \ge 1-\alpha.$$

Now it follows that the set of all imputations which are dominated by some core-element through the coalition {1,2} and {1,3} respectively is equal to

$$\{y \in I(v) \mid y_2 < 1-\alpha < y_3\} \quad \text{and} \quad \{y \in I(v) \mid y_3 < 1-\alpha < y_2\}.$$

We obtain dom $C(v) = I(v) - W$ where the set $W$ is given by

$W := C(v) \cup \text{conv}\,\{(0,\alpha,1-\alpha),\ (0,1-\alpha,\alpha),\ (2\alpha-1,1-\alpha,1-\alpha)\}$.

Here $W = C(v)$ iff $\alpha = 0.5$. Thus, the core is the unique stable set whenever $\alpha = 0.5$. In case $0.5 < \alpha \leq 1$, the game possesses an immense collection of stable sets of the form

$$C(v) \cup \text{conv}\,\{(0,\beta,1-\beta),\ (2\alpha-1,1-\alpha,1-\alpha)\}$$

where $\beta$ is any real number satisfying $1-\alpha \leq \beta \leq \alpha$.

## 5. The bargaining set $\mathcal{M}$

The concepts which have been treated in the previous sections neglect the bargaining process that may actually take place during a play of the game. The internal and external stability conditions for stable sets can be regarded as requirements of behaviour, while the Shapley value meets a certain expectation principle. The various bargaining sets, introduced in Aumann and Maschler (1964), are more closely tied to the bargaining process since they take account of the possible threats and counterthreats made by coalitions. In this work we merely pay attention to the bargaining set which is obtained by taking account of objections and counterobjections made by single players. Here we recall that the set $\Gamma_{ij}$ consists of coalitions containing player i but not player j, on the understanding that $i \neq j$.

DEFINITION 5.1. Let $v \in G^n$ and $x \in I(v)$. An <u>objection</u> of player i against another player j with respect to the imputation x in the game v is a pair $(\hat{y};S)$ where $S \in \Gamma_{ij}$ and $\hat{y} = (y_k)_{k\in S}$ is a $|S|$-tuple of real numbers satisfying

$$\sum_{k\in S} y_k = v(S) \qquad \text{and} \qquad y_k > x_k \text{ for } k \in S. \tag{2.11}$$

A <u>counterobjection</u> to the above objection $(\hat{y};S)$ is a pair $(\hat{z};T)$ where $T \in \Gamma_{ji}$ and $\hat{z} = (z_k)_{k\in T}$ is a $|T|$-tuple of real numbers satisfying

$$\sum_{k\in T} z_k = v(T), \qquad z_k \geq x_k \text{ for } k \in T \qquad \text{and}$$
$$z_k \geq y_k \text{ for } k \in T \cap S. \tag{2.12}$$

Thus, an objection of i against j at an imputation consists of a coalition S containing player i but not player j, and a feasible payoff vector for S that is preferred to the given imputation by every member of the coalition S. Note that a coalition $S \in \Gamma_{ij}$ can be used for an objection of i against j with respect to the imputation x if and only if the corresponding excess $e^v(S,x)$ is positive.
A counterobjection to this objection consists of another coalition T containing player j but not player i, and a feasible payoff vector for T that is weakly preferred to the above payoff vector for S by every member of $T \cap S$ and that is also weakly preferred to the given imputation by every member of T–S. Notice that the excess $e^v(T,x)$ of any coalition T which is used for the counterobjection, must be nonnegative.

DEFINITION 5.2. Let $v \in G^n$. An imputation $x \in I(v)$ is said to belong to the <u>bargaining set</u> $\mathcal{M}(v)$ of the game v if for any objection of one player against another player with respect to the imputation x in the game v, there exists a counterobjection.

Throughout this work we write $\mathcal{M}(v)$ or $\mathcal{M}$ instead of the conventional notation $\mathcal{M}_1^{(i)}(v)$. Since the core of a game consists of imputations with merely nonpositive excesses, there are no objections with respect to any core-element possible. Consequently, the core is always included in the bargaining set $\mathcal{M}$. Thus,

$$C(v) \subset \mathcal{M}(v) \qquad \text{for all } v \in G^n.$$

As a matter of fact, the bargaining set $\mathcal{M}$ of any game is nonempty. Davis and Maschler (1963) presented a direct proof of this statement by considering some properties of the bargaining set $\mathcal{M}$, whereas Peleg (1963, 1967) proved this statement in an indirect way by using Brouwer's fixed point theorem. An alternative proof of the nonemptiness of the bargaining set $\mathcal{M}$ (and a related concept) will be given in Section 7 by specifying a point in the set.

The bargaining set $\mathcal{M}$ is a finite union of closed convex polyhedra. This result is due to Maschler (1966) who described the system of inequalities that determines those polyhedra in explicit form. By this result, the bargaining set $\mathcal{M}$ itself is also closed, but it is in general not a convex set.

EXAMPLE 5.3. Consider again the 3-person game v of (2.10). We assert that the bargaining set $\mathcal{M}$ coincides with the core, i.e., $\mathcal{M}(v) = C(v)$. For that purpose, it is sufficient to show that $x \in I(v) - C(v)$ implies $x \notin \mathcal{M}(v)$.
Let $x \in I(v) - C(v)$. Then $x_2 > 1-\alpha$ or $x_3 > 1-\alpha$. We treat the case $x_2 > 1-\alpha$. Then $e(\{1,3\},x) > 0$ and hence, player 1 can object against player 2 with respect to the imputation x by using the coalition $\{1,3\}$. However, there exists no counterobjection to this objection since $e(\{2\},x) < 0$ and $e(\{2,3\},x) < 0$. So, $x \notin \mathcal{M}(v)$. The case $x \in I(v) - C(v)$ where $x_3 > 1-\alpha$ can be treated in a similar way and is left to the reader. We conclude that $\mathcal{M}(v) = C(v)$.

## 6. The kernel and the prekernel

Closely related to the bargaining set $\mathcal{M}$ is the kernel which was introduced in Davis and Maschler (1965). There it was established that the kernel is a specific subset of the bargaining set $\mathcal{M}$ and thus, it is easier to compute than the bargaining set although even the computation of the kernel may be tedious. The prekernel can be seen as a simplification of the kernel and was introduced in Maschler, Peleg and Shapley (1972) in order to determine the kernel for a certain type of games. The kernel and the prekernel are based on the ideas of excess and maximum surplus.

DEFINITION 6.1. Let $v \in G^n$ and $x \in I^*(v)$.
The <u>maximum</u> <u>surplus</u> of player i over another player j with respect to the pre-imputation x in the game v is given by

$$s_{ij}^v(x) := \max\left(e^v(S,x) \mid S \in \Gamma_{ij}\right). \tag{2.13}$$

A nonnegative (nonpositive respectively) maximum surplus of i over j at a pre-imputation x represents the maximal (minimal) amount what player i can gain (lose) without the cooperation of player j by withdrawing from the payoff vector x and forming a coalition not containing player j, on the understanding that the other members of the formed coalition are satisfied with the amount they had according to the payoff vector x. Thus, the maximum surplus $s_{ij}^v(x)$ can be regarded as a measure of the power of player i to threaten player j with respect to the pre-imputation x.

In case the pre-imputation x is also individually rational, player j is immune to threats whenever $x_j = v(\{j\})$ because player j can get the amount $v(\{j\})$ by going alone. We say player i outweighs player j with respect to the imputation $x \in I(v)$ if

$$x_j > v(\{j\}) \qquad \text{and} \qquad s^v_{ij}(x) > s^v_{ji}(x).$$

The kernel is defined as the set of all imputations for which no player outweighs another player. The prekernel consists of pre-imputations for which any two players are equally powerful concerning their mutual threats. Formally, both the kernel and the prekernel are described as follows.

DEFINITION 6.2. Let $v \in G^n$.
The <u>kernel</u> $\mathcal{K}(v)$ of the game v is the set of all imputations $x \in I(v)$ satisfying for all $i,j \in N$, $i \neq j$,

$$\left(s^v_{ij}(x) - s^v_{ji}(x)\right)\left(x_j - v(\{j\})\right) \le 0 \qquad \text{and}$$

$$\left(s^v_{ji}(x) - s^v_{ij}(x)\right)\left(x_i - v(\{i\})\right) \le 0. \tag{2.14}$$

The <u>prekernel</u> $\mathcal{K}^*(v)$ of the game v is the set of all pre-imputations $x \in I^*(v)$ satisfying

$$s^v_{ij}(x) = s^v_{ji}(x) \qquad \text{for all } i,j \in N,\ i \neq j. \tag{2.15}$$

By the above definition, the kernel is a finite union of closed convex polyhedra because it is determined by a system of inequalities. The relevant polyhedra are studied in Maschler and Peleg (1966) in order to give an algebraic existence proof of the kernel, whereas Davis and Maschler (1965) had already presented an indirect existence proof of the kernel by using Brouwer's fixed point theorem. An alternative proof of the nonemptiness of the kernel (and in addition, the bargaining set $\mathcal{M}$) will be given in Section 7 by specifying a point in the set. The prekernel was treated for the first time in Maschler, Peleg and Shapley (1972) who proved the nonemptiness of the prekernel by adapting the algebraic existence proof of the kernel as given in Maschler and Peleg (1966). We omit the existence proof of the prekernel.

We first prove the announced result that the kernel is a subset of the bargaining set $\mathcal{M}$. Note that we often write $s_{ij}(x)$ instead of $s^v_{ij}(x)$.

THEOREM 6.3. $\mathcal{K}(v) \subset \mathcal{M}(v)$ for all $v \in G^n$.

PROOF. Let $v \in G^n$ and $x \in \mathcal{K}(v)$. Let $(\hat{y};S)$ be an objection of player i against another player j with respect to the kernel-element x in the game v. We show that there exists a counterobjection to this objection.
We have $S \in \Gamma_{ij}$, $y(S) = v(S)$ and $y_k > x_k$ for all $k \in S$. In case $s_{ij}(x) > s_{ji}(x)$, then it follows directly from (2.14) that $x_j = v(\{j\})$ and hence, $(x_j;\{j\})$ is a counterobjection to the objection $(\hat{y};S)$.
It remains to consider the case $s_{ij}(x) \le s_{ji}(x)$. Choose $T \in \Gamma_{ji}$ such that $e(T,x) = s_{ji}(x)$. Now it follows that

$$y(T \cap S) = y(S) - y(S-T) = v(S) - y(S-T)$$
$$< v(S) - x(S-T) = v(S) - x(S) + x(S \cap T) \quad \text{and hence,}$$

$$y(T \cap S) + x(T-S) < v(S) - x(S) + x(T)$$
$$\le s_{ij}(x) + x(T) = s_{ij}(x) - s_{ji}(x) + v(T) \le v(T).$$

In view of (2.12), the inequality $v(T) > y(T \cap S) + x(T-S)$ implies the existence of a counterobjection to the above objection $(\hat{y};S)$ by using the coalition T.
From the above we conclude that $x \in \mathcal{M}(v)$. □

Clearly, $\mathcal{K}^*(v) \cap I(v) \subset \mathcal{K}(v)$. Since the prekernel may contain pre-imputations that are not individually rational, the prekernel is in general not included in the bargaining set $\mathcal{M}$ and hence, it is usually different from the kernel. In Section V.7 we shall show that the kernel and the prekernel coincide for a large class of games, including the superadditive games. As a matter of fact, the parts of the kernel and the prekernel inside the core always coincide. This one statement as well as the subsequent statements concerning the geometric structure of the (pre)kernel are due to Maschler, Peleg and Shapley (1979).

THEOREM 6.4. $\mathcal{K}(v) \cap C_\epsilon(v) = \mathcal{K}^*(v) \cap C_\epsilon(v)$ for all $v \in G^n$ and all $\epsilon \le 0$.

PROOF. Let $v \in G^n$ and $\epsilon \le 0$. Because $C_\epsilon(v) \subset I(v)$ as well as $\mathcal{K}^*(v) \cap I(v) \subset \mathcal{K}(v)$, it is obvious that

$$\mathcal{K}^*(v) \cap C_\epsilon(v) \subset \mathcal{K}(v) \cap C_\epsilon(v).$$

To prove the inverse inclusion, suppose $x \in \mathcal{K}(v) \cap C_\epsilon(v)$. We must show that $x \in \mathcal{K}^*(v)$. Assume $x \notin \mathcal{K}^*(v)$. Then there exist $i \in N$, $j \in N$, $i \neq j$, such that $s_{ij}(x) \neq s_{ji}(x)$, say $s_{ij}(x) > s_{ji}(x)$. Then $x_j = v(\{j\})$ by (2.14) and therefore,

$$s_{ij}(x) > s_{ji}(x) \geq v(\{j\}) - x_j = 0 \geq \epsilon.$$

Choose $T \in \Gamma_{ij}$ such that $e(T,x) = s_{ij}(x)$. Now it follows that $v(T) - x(T) = e(T,x) = s_{ij}(x) > \epsilon$, but this strict inequality is in contradiction with $x \in C_\epsilon(v)$ by (2.7). So, $x \in \mathcal{K}(v) \cap C_\epsilon(v)$ implies $x \in \mathcal{K}^*(v) \cap C_\epsilon(v)$. □

Now we look at the geometric characterization for the intersections of the (pre)kernel with the strong $\epsilon$-cores. For any $i \in N$, let $e^i \in \mathbb{R}^n$ denote the i-th unit vector.

DEFINITION 6.5. Let $v \in G^n$, $\epsilon \in \mathbb{R}$, $x \in C_\epsilon(v)$ and $i,j \in N$, $i \neq j$. The critical number $\delta^v_{ij}(\epsilon,x) \in \mathbb{R}$ is defined by

$$\delta^v_{ij}(\epsilon,x) := \max\left[\delta \mid \delta \in \mathbb{R},\ x - \delta e^i + \delta e^j \in C_\epsilon(v)\right]. \tag{2.16}$$

In addition, let $R^v_{ij}(\epsilon,x)$ be the straight line segment with end points

$x - \delta^v_{ij}(\epsilon,x)e^i + \delta^v_{ij}(\epsilon,x)e^j$ and $x + \delta^v_{ji}(\epsilon,x)e^i - \delta^v_{ji}(\epsilon,x)e^j$.

The nonnegative real number $\delta^v_{ij}(\epsilon,x)$ represents the largest amount that can be transferred from player i to player j with respect to the strong $\epsilon$-core-element x while remaining in the strong $\epsilon$-core of the game v. This critical number is well-defined since any strong $\epsilon$-core of a game $v \in G^n$ is a compact convex subset of $\mathbb{R}^n$. So, $x - \delta^v_{ij}(\epsilon,x)e^i + \delta^v_{ij}(\epsilon,x)e^j$ is the unique boundary point of the strong $\epsilon$-core that is obtained from the point x by letting its i-th coordinate decrease and its j-th coordinate increase by the same nonnegative amount. Consequently, the line segment $R^v_{ij}(\epsilon,x)$ represents the largest straight line segment through the point x in the i-j direction that is still included in the strong $\epsilon$-core. Thus, it is clear that $R^v_{ij}(\epsilon,x) = R^v_{ji}(\epsilon,x)$.

According to the next lemma, the critical number of (2.16) is closely related to the maximum surplus. Subsequently, the geometric characterization for the intersection of the (pre)kernel with a nonempty strong $\epsilon$-core is formulated in terms of a bisection property for the various straight line segments induced by any strong $\epsilon$-core-element.

LEMMA 6.6. Let $v \in G^n$, $\epsilon \in \mathbb{R}$ and $x \in C_\epsilon(v)$. Then

$$\delta^v_{ij}(\epsilon,x) = \epsilon - s^v_{ij}(x) \qquad \text{for all } i,j \in N,\ i \neq j.$$

PROOF. Let $i,j \in N$, $i \neq j$, and put $x^\delta := x - \delta e^i + \delta e^j \in \mathbb{R}^n$ for any $\delta \in \mathbb{R}$. By (1.6), we have that for all $S \subset N$

$$e(S,x^\delta) = v(S) - x^\delta(S),\ e(S,x) = v(S) - x(S) \quad \text{and hence,}$$

$$\begin{aligned} e(S,x^\delta) &= e(S,x) + \delta && \text{if } S \in \Gamma_{ij} \\ &= e(S,x) - \delta && \text{if } S \in \Gamma_{ji} \\ &= e(S,x) && \text{otherwise.} \end{aligned}$$

Further, $e(S,x) \leq \epsilon$ for all $S \neq N,\emptyset$ because of $x \in C_\epsilon(v)$. Now it follows that for any $\delta \geq 0$

$$\begin{aligned} x^\delta \in C_\epsilon(v) \quad &\text{iff} \quad e(S,x^\delta) \leq \epsilon \text{ for all } S \neq N,\emptyset \\ &\text{iff} \quad \max\left(e(S,x) \mid S \in \Gamma_{ij}\right) + \delta \leq \epsilon \\ &\text{iff} \quad \delta \leq \epsilon - s_{ij}(x). \end{aligned}$$

From this equivalence and (2.16), we immediately conclude that $\delta_{ij}(\epsilon,x) = \epsilon - s_{ij}(x)$. □

THEOREM 6.7. Let $v \in G^n$, $\epsilon \in \mathbb{R}$ and $x \in C_\epsilon(v)$.

(i) Then $x \in \mathcal{K}^*(v) \cap C_\epsilon(v)$ iff $x$ bisects the straight line segment $R^v_{ij}(\epsilon,x)$ for all $i,j \in N$, $i \neq j$.

(ii) Then $x \in \mathcal{K}(v) \cap C_\epsilon(v)$ iff $x \in I(v)$ and for all $i,j \in N$, $i \neq j$,
either $x$ bisects the straight line segment $R^v_{ij}(\epsilon,x)$
or $\left[x_i = v(\{i\}) \text{ and } \delta^v_{ij}(\epsilon,x) > \delta^v_{ji}(\epsilon,x)\right]$
or $\left[x_j = v(\{j\}) \text{ and } \delta^v_{ij}(\epsilon,x) < \delta^v_{ji}(\epsilon,x)\right]$.

(iii) In addition, let $\epsilon = 0$.
Then $x \in \mathcal{K}(v) \cap C(v)$ iff $x$ bisects the straight line segment $R^v_{ij}(0,x)$ for all $i,j \in N$, $i \neq j$.

PROOF. (i) It follows directly from (2.15) and Lemma 6.6 respectively that

$$x \in \mathcal{K}^*(v) \quad \text{iff} \quad s_{ij}(x) = s_{ji}(x) \quad \text{for all } i,j \in N,\ i \neq j$$
$$\text{iff} \quad \delta_{ij}(\epsilon,x) = \delta_{ji}(\epsilon,x) \text{ for all } i,j \in N,\ i \neq j.$$

(ii) From (2.14) we deduce that $x \in \mathcal{K}(v)$ iff $x \in I(v)$ and for all $i,j \in N$, $i \neq j$,

$$\text{either } s_{ij}(x) = s_{ji}(x)$$
$$\text{or } [x_i = v(\{i\}) \text{ and } s_{ij}(x) < s_{ji}(x)]$$
$$\text{or } [x_j = v(\{j\}) \text{ and } s_{ij}(x) > s_{ji}(x)].$$

In view of Lemma 6.6, the statement (ii) becomes clear. The statement (iii) is a direct consequence of part (i) and Theorem 6.4 applied to $\epsilon = 0$. □

Here an imputation $x \in I(v)$ is said to be interior to the imputation set $I(v)$ whenever $x_i > v(\{i\})$ for all $i \in N$. The bisection property for the intersection of the prekernel (kernel respectively) with a nonempty strong $\epsilon$-core states that a strong $\epsilon$-core-element which is interior to the imputation set belongs to the prekernel (kernel) if and only if the involved point of the strong $\epsilon$-core is always the midpoint of the largest straight line segments through the involved point in the various directions that are still included in the strong $\epsilon$-core.
The bisection property for the intersection of the kernel with the nonempty core expresses that the intersection merely depends upon the geometric structure of the core. Particularly, for any two games $v, w \in G^n$ with identical cores we have the equality

$$\delta^v_{ij}(0,x) = \delta^w_{ij}(0,x) \quad \text{and hence,} \quad R^v_{ij}(0,x) = R^w_{ij}(0,x)$$

for all $x \in C(v)$ and all $i \in N$, $j \in N$, $i \neq j$. Together with Theorem 6.7(iii), this reasoning proves the next statement.

COROLLARY 6.8. If $v, w \in G^n$ are such that $C(v) = C(w)$, then

$$\mathcal{K}(v) \cap C(v) = \mathcal{K}(w) \cap C(w).$$

The above corollary is of the highest importance in order to prove in Section VII.7 that for so-called k-convex games the kernel intersects the core at a unique point.

Before we illustrate the bisection properties of the (pre)kernel by means of an example, we also establish two other useful properties of the (pre)kernel which are due to Maschler and Peleg (1966).

THEOREM 6.9. Let $v \in G^n$ and $x \in \mathcal{K}(v) \cup \mathcal{K}^*(v)$.

(i) Substitution property of the (pre)kernel. Let the players i and j be substitutes in the game v, i.e.,

$$v(S \cup \{i\}) = v(S \cup \{j\}) \quad \text{for all } S \subset N-\{i,j\}. \qquad (2.17)$$

Then $x_i = x_j$.

(ii) Let $i \in N$ be a dummy in the game v. Then $x_i \leq v(\{i\})$. In particular, the kernel possesses the dummy player property: $x_i = v(\{i\})$ whenever $x \in \mathcal{K}(v)$.

PROOF. (i) Let $i \in N$ and $j \in N$ be such that (2.17) holds. Assume $x_i \neq x_j$, say $x_i > x_j$. Choose a coalition $S \in \Gamma_{ij}$ such that $e(S,x) = s_{ij}(x)$. Put $T := (S-\{i\}) \cup \{j\}$. Then we have $v(T) = v(S)$ by using (2.17). Now it follows that

$$\begin{aligned} s_{ji}(x) &\geq e(T,x) = v(T) - x(T) = v(S) - x(S) + x_i - x_j \\ &> v(S) - x(S) = e(S,x) = s_{ij}(x). \end{aligned}$$

From the strict inequality $s_{ji}(x) > s_{ij}(x)$ we deduce a contradiction with (2.15) when $x \in \mathcal{K}^*(v)$. In case $x \in \mathcal{K}(v)$, then we obtain a contradiction with (2.14) by noting that $s_{ji}(x) > s_{ij}(x)$ and $x_i > x_j \geq v(\{j\}) = v(\{i\})$, where the equality follows from (2.17). We conclude that $x_i = x_j$.

(ii) Let player $i \in N$ be a dummy in the game v. Assume $x_i > v(\{i\})$. Choose $j \in N-\{i\}$ such that $s_{ji}(x) \geq s_{ki}(x)$ for all $k \in N-\{i\}$. We show that $s_{ji}(x) > s_{ij}(x)$ in order to obtain a contradiction with both (2.14) and (2.15). We have $s_{ji}(x) > 0$ because

$$\begin{aligned} s_{ji}(x) &\geq e(N-\{i\},x) = v(N-\{i\}) - x(N-\{i\}) \\ &= v(N) - v(\{i\}) - x(N) + x_i = x_i - v(\{i\}) > 0. \end{aligned}$$

Choose $S \in \Gamma_{ij}$ such that $s_{ij}(x) = e(S,x)$. In case $S = \{i\}$, then $s_{ij}(x) = v(\{i\}) - x_i < 0 < s_{ji}(x)$. So, it remains to consider the case $S \neq \{i\}$. Then there exists $k \in S-\{i\}$. Now it follows that

$$\begin{aligned} s_{ij}(x) &= v(S) - x(S) = v(S-\{i\}) + v(\{i\}) - x(S) \\ &= e(S-\{i\},x) + v(\{i\}) - x_i \\ &< e(S-\{i\},x) \le s_{ki}(x) \le s_{ji}(x). \end{aligned}$$ □

Two players are called substitutes whenever the worth of any coalition containing only one of both players is not affected by replacing that one player by the other player. According to any payoff vector in the (pre)kernel, substitutes receive the same payoff. The game v of Example 4.4 illustrates that the core does not possess the substitution property because the substitutes 2 and 3 receive different payoffs with respect to the core-element $(\alpha,1-\alpha,0)$ whenever $0.5 \le \alpha < 1$. It is left to the reader to derive from the familiar formula (2.3) by straightforward calculations that the Shapley value also possesses the substitution property.

EXAMPLE 6.10. Consider once again the 3-person game v of (2.10). Because the players 2 and 3 are substitutes in the game v, it follows from Theorem 6.9 that the (pre)kernel of the game v only contains pre-imputations of the form $x = x(\beta) = (1-2\beta,\beta,\beta)$ where $\beta \in \mathbb{R}$. Then, by using (2.13) as well as (1.6), we have

$$\begin{aligned} & & s_{23}(x) &= s_{32}(x), \\ s_{12}(x) &= \max[2\beta-1,\ \alpha-1+\beta], & s_{13}(x) &= s_{12}(x), \\ s_{21}(x) &= \max[-\beta,\ -2\beta], & s_{31}(x) &= s_{21}(x). \end{aligned}$$

So, $$s_{12}(x) - s_{21}(x) \begin{cases} > 0 & \text{if } \beta > \frac{1}{2}(1-\alpha) \\ < 0 & \text{if } \beta < \frac{1}{2}(1-\alpha) \\ = 0 & \text{if } \beta = \frac{1}{2}(1-\alpha). \end{cases}$$

By straightforward calculations, we obtain from (2.14) and (2.15) that $\mathcal{K}(v) = \mathcal{K}^*(v) = \{\frac{1}{2}(2\alpha,1-\alpha,1-\alpha)\}$.

In order to illustrate the bisection properties of the (pre)kernel, we look at various strong $\epsilon$-cores of the game where $\alpha = 0.75$. Each strong $\epsilon$-core of the game v is bounded by the hyperplanes $H_S^\epsilon$, $S \neq N,\emptyset$, given by

$$H_S^\epsilon := \{x \in I^*(v) \mid x(S) = v(S) - \epsilon\} \qquad \text{for all } S \neq N,\emptyset.$$

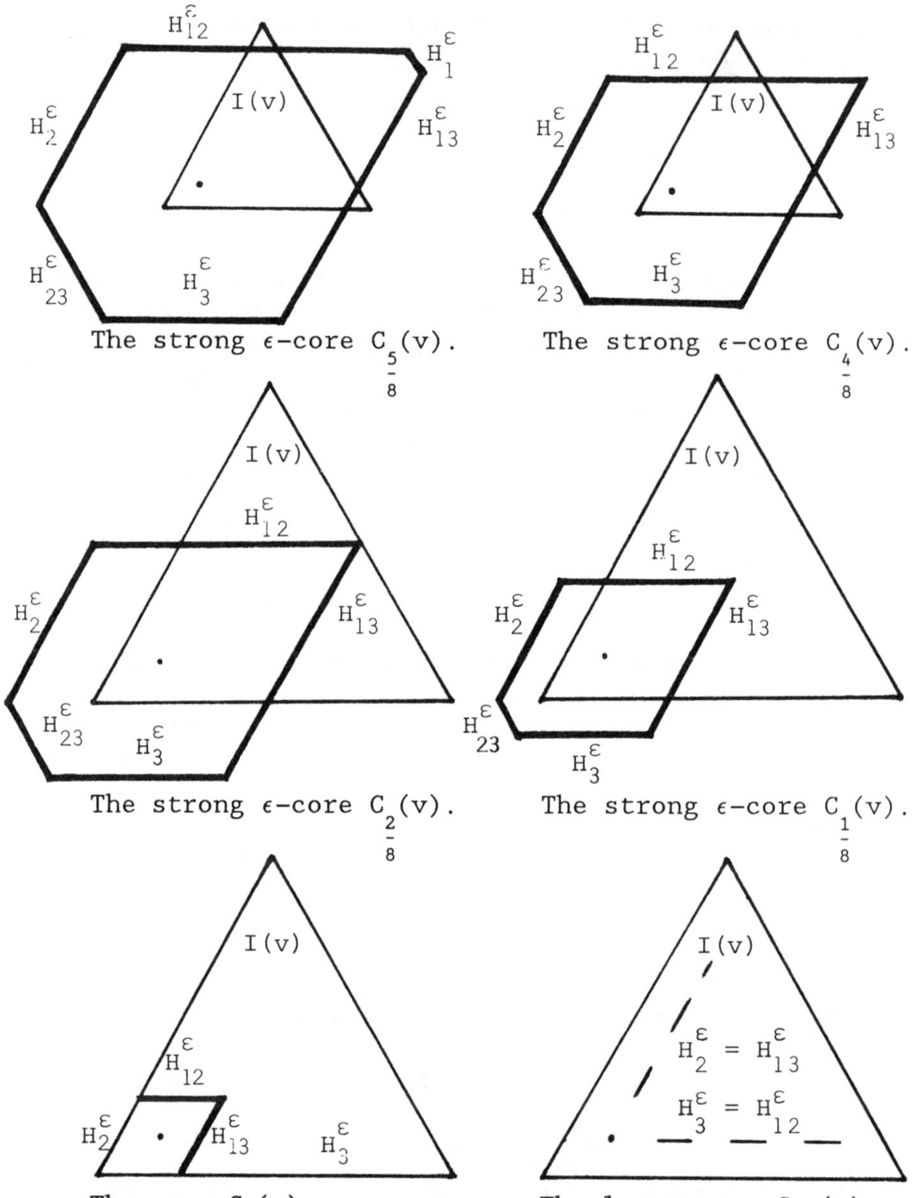


The strong $\epsilon$-core $C_{\frac{5}{8}}(v)$.

The strong $\epsilon$-core $C_{\frac{4}{8}}(v)$.

The strong $\epsilon$-core $C_{\frac{2}{8}}(v)$.

The strong $\epsilon$-core $C_{\frac{1}{8}}(v)$.

The core $C_0(v)$.

The least-core $C_{-\frac{1}{8}}(v)$.

FIGURE 2.2. The strong $\epsilon$-cores $C_\epsilon(v)$ of the 3-person game v of (2.10) where $\epsilon \in \{\frac{5}{8}, \frac{4}{8}, \frac{2}{8}, \frac{1}{8}, 0, -\frac{1}{8}\}$ and $\alpha = 0.75$. The indicated point inside the strong $\epsilon$-cores represents the (pre)kernel of the game.

For $\epsilon = \frac{5}{8}, \frac{1}{2}, \frac{1}{4}, \frac{1}{8}$, 0 and $-\frac{1}{8}$ respectively, the position of the strong $\epsilon$-core with respect to the imputation set I(v) is drawn in Figure 2.2. Although these six strong $\epsilon$-cores differ in shape, the unique (pre)kernel-element $(\frac{3}{4},\frac{1}{8},\frac{1}{8})$ always bisects the three straight line segments through the involved point in the various directions that are still included in the strong $\epsilon$-core. The reader is invited to verify this in Figure 2.2.

EXAMPLE 6.11. We consider an 8-person game whose player set N is partitioned into two distinct subsets $P = \{1,2,3,4\}$ and $Q = \{5,6,7,8\}$. Let the characteristic function v be given by

$$v(N) = 40, \quad v(N-\{i\}) = 34 \text{ for all } i \in N,$$

$$v\big((P \cup \{q\})-\{p\}\big) = \alpha \qquad \text{for all } p \in P \text{ and all } q \in Q,$$

$$v\big((Q \cup \{p\})-\{q\}\big) = \alpha \qquad \text{for all } p \in P \text{ and all } q \in Q,$$

$$v(S) = 0 \qquad \text{otherwise}$$

where $\alpha$ is any real number satisfying $\alpha \geq 0$.
From the substitution property of the (pre)kernel applied to the members of P and Q respectively in the game v, we deduce that the (pre)kernel of the game v only contains pre-imputations of the form

$$x = x(\beta) = (\beta,\ \beta,\ \beta,\ \beta,\ 10-\beta,\ 10-\beta,\ 10-\beta,\ 10-\beta)$$

where $\beta \in \mathbb{R}$. Evidently, $x(\beta) \in I(v)$ iff $0 \leq \beta \leq 10$. The maximum surplus is determined with the aid of (2.13) as well as (1.6). By straightforward calculations, we obtain that for all $p \in P$ and all $q \in Q$

$$s_{pq}(x) = \max\big(-4\beta,\ -2\beta-10+\alpha,\ -\beta+4,\ \beta-20,\ 2\beta-30+\alpha\big),$$

$$s_{qp}(x) = \max\big(-2\beta-10+\alpha,\ -\beta-10,\ \beta-6,\ 2\beta-30+\alpha,\ 4\beta-40\big).$$

The above general formula for the maximum surplus can be reduced in a geometric way to the detailed form that is listed in Table 2.3. Especially, we are interested in the positive or negative sign of the difference $s_{pq}(x) - s_{qp}(x)$ for any $p \in P$ and any $q \in Q$. From Table 2.3, (2.14) and (2.15), we conclude the following results concerning the kernel and the prekernel of the game v.

| The case $\frac{38}{3} \le \alpha < 19$ | | | |
|---|---|---|---|
| $\beta$ | $s_{pq}(x)$ | $s_{qp}(x)$ | $\beta$ |
| $\beta \le 5-\frac{\alpha}{2}$ | $-4\beta$ | $-2\beta-10+\alpha$ | $\beta \le \frac{1}{3}(\alpha-4)$ |
| $5-\frac{\alpha}{2} \le \beta \le \alpha-14$ | $-2\beta-10+\alpha$ | $\beta-6$ | $\frac{1}{3}(\alpha-4) \le \beta \le 24-\alpha$ |
| $\alpha-14 \le \beta \le \frac{1}{3}(34-\alpha)$ | $-\beta+4$ | $2\beta-30+\alpha$ | $24-\alpha \le \beta \le 5+\frac{\alpha}{2}$ |
| $\beta \ge \frac{1}{3}(34-\alpha)$ | $2\beta-30+\alpha$ | $4\beta-40$ | $\beta \ge 5+\frac{\alpha}{2}$ |

| The case $\alpha \ge 19$ | | | |
|---|---|---|---|
| $\beta$ | $s_{pq}(x)$ | $s_{qp}(x)$ | $\beta$ |
| $\beta \le 5-\frac{\alpha}{2}$ | $-4\beta$ | $-2\beta-10+\alpha$ | $\beta \le 5$ |
| $5-\frac{\alpha}{2} \le \beta \le 5$ | $-2\beta-10+\alpha$ | $2\beta-30+\alpha$ | $5 \le \beta \le 5+\frac{\alpha}{2}$ |
| $\beta \ge 5$ | $2\beta-30+\alpha$ | $4\beta-40$ | $\beta \ge 5+\frac{\alpha}{2}$ |

| The case $0 \le \alpha < \frac{38}{3}$ | | | |
|---|---|---|---|
| $\beta$ | $s_{pq}(x)$ | $s_{qp}(x)$ | $\beta$ |
| $\beta \le -\frac{4}{3}$ | $-4\beta$ | $-2\beta-10+\alpha$ | $\beta \le \frac{1}{3}(\alpha-4)$ |
| $-\frac{4}{3} \le \beta \le \frac{1}{3}(34-\alpha)$ | $-\beta+4$ | $\beta-6$ | $\frac{1}{3}(\alpha-4) \le \beta \le \frac{34}{3}$ |
| $\beta \ge \frac{1}{3}(34-\alpha)$ | $2\beta-30+\alpha$ | $4\beta-40$ | $\beta \ge \frac{34}{3}$ |

| The case $\frac{38}{3} \le \alpha < 19$ | | | The case $\alpha \ge 19$ |
|---|---|---|---|
| $\beta$ | $s_{pq}(x) - s_{qp}(x)$ | | $\beta$ |
| $\beta < 5-\frac{\alpha}{2}$ | $> 0$ | $> 0$ | $\beta < 5-\frac{\alpha}{2}$ |
| $5-\frac{\alpha}{2} \le \beta \le \alpha-14$ | $= 0$ | $= 0$ | $5-\frac{\alpha}{2} \le \beta \le 5+\frac{\alpha}{2}$ |
| $\alpha-14 < \beta < 5$ | $> 0$ | $< 0$ | $\beta > 5+\frac{\alpha}{2}$ |
| $\beta = 5$ | $= 0$ | The case $0 \le \alpha < \frac{38}{3}$ | |
| $5 < \beta < 24-\alpha$ | $< 0$ | $> 0$ | $\beta < 5$ |
| $24-\alpha \le \beta \le 5+\frac{\alpha}{2}$ | $= 0$ | $= 0$ | $\beta = 5$ |
| $\beta > 5+\frac{\alpha}{2}$ | $< 0$ | $< 0$ | $\beta > 5$ |

TABLE 2.3. The maximum surplus $s_{pq}(x)$ and $s_{qp}(x)$ in the game $v$ of Example 6.11 where $p \in P$, $q \in Q$ and $x = x(\beta) = (\beta, \beta, \beta, \beta, 10-\beta, 10-\beta, 10-\beta, 10-\beta)$.

$$
\begin{aligned}
\mathcal{K}^*(v) &= \{x(\beta) \mid 5-\tfrac{\alpha}{2} \le \beta \le 5+\tfrac{\alpha}{2}\} && \text{if } \alpha \ge 19 \\
&= \{x(\beta) \mid 5-\tfrac{\alpha}{2} \le \beta \le \alpha-14 \text{ or} \\
&\qquad \beta = 5 \text{ or } 24-\alpha \le \beta \le 5+\tfrac{\alpha}{2}\} && \text{if } \tfrac{38}{3} \le \alpha < 19 \\
&= \{x(\beta) \mid \beta = 5\} && \text{if } 0 \le \alpha < \tfrac{38}{3} \\
\mathcal{K}(v) &= \{x(\beta) \mid 0 \le \beta \le 10\} && \text{if } \alpha \ge 19 \\
&= \{x(\beta) \mid 0 \le \beta \le \alpha-14 \text{ or} \\
&\qquad \beta = 5 \text{ or } 24-\alpha \le \beta \le 10\} && \text{if } 14 \le \alpha < 19 \\
&= \{x(\beta) \mid \beta = 5\} && \text{if } 0 \le \alpha < 14.
\end{aligned}
$$

In particular, $\mathcal{K}(v) = \mathcal{K}^*(v) \cap I(v)$ and as such, the kernel of the game v where $\alpha \ge 38/3$ is strictly included in the prekernel because a part of the prekernel lies outside the imputation set $I(v)$.
If the nonnegative real number $\alpha$ is large enough (i.e., $\alpha \ge 19$), then the (pre)kernel of the game v consists of one straight line segment. If the number $\alpha$ is neither too large nor too small, then the (pre)kernel of v consists of two disconnected straight line segments and one single point. Notice that the kernel (prekernel respectively) degenerates into three points whenever $\alpha = 14$ ($\alpha = 38/3$). If $\alpha$ is small enough, then the (pre)kernel is a singleton. The imputation $x(\beta)$ where $\beta = 5$ occupies a central position within the kernel and in fact, it represents the so-called nucleolus of the game v.

## 7. The nucleolus

An algebraic proof of the nonemptiness of the kernel has been presented in Maschler and Peleg (1966). The same method of proof was used there to show that the kernel $\mathcal{K}(v)$ of a game v intersects any nonempty set $C_\epsilon(v) \cap I(v)$ which represents the part of the strong $\epsilon$-core that is individually rational. Pursuing this result, Schmeidler (1969) was led to the discovery of the nucleolus.

Let $v \in G^n$. For any n-tuple $x \in \mathbb{R}^n$, let $\theta(x)$ be the $2^n$-tuple whose components are the excesses $e^v(S,x)$, $S \subset N$, arranged in nonincreasing order. Thus,

$$\theta_i(x) \ge \theta_j(x) \qquad \text{whenever } 1 \le i \le j \le 2^n.$$

The involved excesses are usually nonpositive (e.g., at core-elements) and therefore, the excesses are regarded as losses or complaints, while the vectors $\theta(x)$, $x \in \mathbb{R}^n$, are interpreted as complaint vectors.
Now the lexicographic order $\leq_L$ on $\mathbb{R}^{2^n}$ is used to order the complaint vectors by taking into account their largest complaint or, if such should be the case, their second largest complaint and so on. For all $x, y \in \mathbb{R}^n$ we write

$\theta(x) <_L \theta(y)$ if there exists an integer $1 \leq k \leq 2^n$ such that $\theta_i(x) = \theta_i(y)$ for $1 \leq i < k$, whereas $\theta_k(x) < \theta_k(y)$;

$\theta(x) \leq_L \theta(y)$ if either $\theta(x) = \theta(y)$ or $\theta(x) <_L \theta(y)$.

DEFINITION 7.1. The <u>nucleolus</u> $\mathcal{N}(v)$ of a game $v \in G^n$ is the set of all imputations $x \in I(v)$ satisfying

$$\theta(x) \leq_L \theta(y) \qquad \text{for all } y \in I(v).$$

Thus, the nucleolus consists of imputations that minimize the complaint function $\theta(x)$ in the lexicographic order over the nonempty compact convex imputation set. Schmeidler (1969) himself gave both a topological and an algebraic existence proof of the nucleolus in terms of continuous functions and nonempty compact sets. Here we treat an alternative existence (and in addition, a uniqueness) proof of the nucleolus concept which is based on the geometric characterization of the nucleolus presented in Maschler, Peleg and Shapley (1979).
Informally, their constructive proof proceeds as follows. First of all, we determine the nonempty compact convex set of all imputations that minimize the maximum excess over the nontrivial coalitions. Then we remove the coalitions whose excess with respect to the imputations of this set can not be further reduced. Secondly, we begin over again to minimize the maximum excess over the remaining coalitions. The resulting compact convex subset of the previous set of imputations is in general nonempty and once again, we remove the coalitions whose excess with respect to the imputations of this second set can not be further lowered. This procedure continues, but it stops whenever all nontrivial coalitions are removed. Formally, the procedure is described as follows.

DEFINITION 7.2. Let $v \in G^n$ where $n \geq 2$. We define

$$X^0 := I(v), \qquad \Sigma^0 := \{S \subset N \mid S \neq N, \emptyset\}$$

and for $j = 1, 2, \cdots, \kappa$, we define recursively

$$\epsilon^j := \min_{x \in X^{j-1}} \max_{S \in \Sigma^{j-1}} e^v(S,x) \tag{2.18}$$

$$X^j := \{x \in X^{j-1} \mid \max_{S \in \Sigma^{j-1}} e^v(S,x) = \epsilon^j\} \tag{2.19}$$

$$\Sigma_j := \{S \in \Sigma^{j-1} \mid e^v(S,x) = \epsilon^j \text{ for all } x \in X^j\} \tag{2.20}$$

$$\Sigma^j := \Sigma^{j-1} - \Sigma_j \tag{2.21}$$

where $$\kappa := \min\big(j \mid j \geq 1,\ \Sigma^j = \emptyset\big). \tag{2.22}$$

Before we establish its correctness, we apply the above procedure to the three-person game v of Example 4.4 where $0.5 < \alpha < 1$. Then $I(v) = \{x \in \mathbb{R}^3_+ \mid x_1 + x_2 + x_3 = 1\}$ and also

$$\epsilon^1 = \min_{x \subset I(v)} \max\big(-x_1,\ -x_2,\ -x_3,\ x_3+\alpha-1,\ x_2+\alpha-1\big).$$

Note that $-x_3 \leq \frac{1}{2}(\alpha-1)$ if and only if $x_3+\alpha-1 \geq \frac{1}{2}(\alpha-1)$. From this equivalence and the formulas (2.18)-(2.22), we obtain

$$\epsilon^1 = \tfrac{1}{2}(\alpha-1), \qquad X^1 = \{\tfrac{1}{2}(2\alpha, 1-\alpha, 1-\alpha)\},$$

$$\Sigma_1 = \{\{2\},\{3\},\{1,2\},\{1,3\}\}, \qquad \Sigma^1 = \{\{1\},\{2,3\}\},$$

$$\epsilon^2 = \max\big(-\alpha,\ \alpha-1\big) = \alpha-1, \qquad X^2 = X^1,$$

$$\Sigma_2 = \{\{2,3\}\}, \qquad \Sigma^2 = \{\{1\}\},$$

$$\epsilon^3 = -\alpha, \qquad X^3 - X^1,$$

$$\Sigma_3 = \{\{1\}\}, \qquad \Sigma^3 = \emptyset \quad \text{and} \quad \kappa = 3.$$

The real numbers $\epsilon^1$, $\epsilon^2$, $\epsilon^3$, and the sets $\Sigma_1$, $\Sigma_2$, $\Sigma_3$ of coalitions can be explained in geometric terms as follows.

We pay attention to various strong $\epsilon$-cores of the game v where $\alpha = 0.75$. In particular, we recall to mind Figure 2.2. The geometric procedure starts with any strong $\epsilon$-core that intersects the imputation set I(v), say $\epsilon = 5/8$. Next we push all the hyperplanes $H_S^{\epsilon}$, $S \neq N,\emptyset$, which are the bounds for the chosen strong $\epsilon$-core. The push is performed at equal speeds and depends upon the shape of the chosen strong $\epsilon$-core as well as its position with respect to the imputation set I(v). We push so long as the enclosed set is neither empty nor disjoint from the imputation set I(v). As a matter of fact, the push is stopped whenever the hyperplanes $H_S^{\epsilon^1}$, $S \neq N,\emptyset$, are met where $\epsilon^1 = -1/8$. Then the enclosed set is equal to $X^1$, being the singleton $\{(\frac{3}{4},\frac{1}{8},\frac{1}{8})\}$. In view of (2.20), the hyperplane $H_S^{\epsilon^1}$ contains the set $X^1$ if and only if $S \in \Sigma_1$. By (2.19), the set $X^1$ would become empty by any further push of a hyperplane $H_S^{\epsilon^1}$ where $S \in \Sigma_1$. Hence, we continue to push only those hyperplanes $H_S^{\epsilon^1}$ where $S \in \Sigma^0 - \Sigma_1 = \{\{1\},\{2,3\}\}$. Once again, we push so long as the enclosed subset of $X^1$ is neither empty nor disjoint from the imputation set. The resulting set is equal to $X^2$. In this second stage we move the two remaining hyperplanes $x_1 = 1/8$ and $x_1 = 7/8$ at the same speed towards the singleton $X^1$. Consequently, the amount of pushing in the second stage is equal to 1/8 and therefore, $\epsilon^2 = -0.25$. The resulting set $X^2$ coincides with the set $X^1$. In the third and final stage, the one remaining hyperplane $x_1 = 0.25$ is moved towards the singleton $X^2$ where the amount of pushing equals 0.5. Therefore, $\epsilon^3 = -0.75$. Now there exist no hyperplanes which can be further pushed and the procedure ends up with the singleton $X^{\kappa} = X^3 = X^2 = X^1$.

The correctness of the procedure is derived from the next two lemmas.

LEMMA 7.3. Let $v \in G^n$ where $n \geq 2$.

(i) The real numbers $\epsilon^j$ are well-defined for $1 \leq j \leq \kappa$.

(ii) The sets $X^j$ are nonempty compact convex subsets of $\mathbb{R}^n$ for $1 \leq j \leq \kappa$ such that $I(v) \supset X^1 \supset X^2 \supset \cdots \supset X^{\kappa}$.

PROOF. The two statements are shown by induction on j, $1 \le j \le \kappa$.
(a) First of all, we prove the induction basis for $j = 1$. We have $\Sigma^0 \ne \emptyset$ since $n > 2$, while the set $X^0$ is equal to the nonempty compact convex imputation set $I(v)$. Let the function $f_0\colon X^0 \to \mathbb{R}$ be given by

$$f_0(x) := \max\{e(S,x) \mid S \in \Sigma^0\} \qquad \text{for all } x \in X^0.$$

Then the function $f_0$ is well-defined. Furthermore, $f_0$ is a continuous function of x since it is defined as maximum of a finite number of continuous functions of x. Therefore, the continuous function $f_0$ on the compact set $X^0$ attains its minimum in a compact subset of $X^0$. By (2.18), (2.19) and the definition of the function $f_0$, we get

$$\epsilon^1 := \min\{f_0(x) \mid x \in X^0\} \qquad \text{and} \qquad X^1 := \{x \in X^0 \mid f_0(x) = \epsilon^1\}.$$

Now it follows that $\epsilon^1$ is well-defined and $X^1 \ne \emptyset$, compact. In order to show that the set $X^1$ is also convex, let $x \in X^1$, $y \in X^1$ and $0 < \alpha < 1$. Then we have $x \in X^0$, $y \in X^0$ and further, $f_0(x) = f_0(y) = \epsilon^1$. Put $z := \alpha x + (1-\alpha)y$. From the convexity of the set $X^0$ we derive that $z \in X^0$ and so, $f_0(z) \ge \epsilon^1$. But we also have

$$\begin{aligned} f_0(z) &= \max\{v(S)-z(S) \mid S \in \Sigma^0\} \\ &= \max\{\alpha[v(S)-x(S)] + (1-\alpha)[v(S)-y(S)] \mid S \in \Sigma^0\} \\ &\le \alpha\max\{v(S)-x(S) \mid S \in \Sigma^0\} + (1-\alpha)\max\{v(S)-y(S) \mid S \in \Sigma^0\} \\ &= \alpha f_0(x) + (1-\alpha)f_0(y) = \alpha\epsilon^1 + (1-\alpha)\epsilon^1 = \epsilon^1. \end{aligned}$$

We conclude that $f_0(z) = \epsilon^1$ or equivalently, $z \in X^1$. Therefore, the set $X^1$ is convex. This completes the proof of the induction basis for $j = 1$.
(b) Let $2 \le j \le \kappa$ and suppose that $X^{j-1}$ is a nonempty compact convex set and $\epsilon^{j-1}$ is well-defined. Then $\Sigma^{j-1} \ne \emptyset$ by (2.22). Let the function $f_{j-1}\colon X^{j-1} \to \mathbb{R}$ be given by

$$f_{j-1}(x) := \max\{e(S,x) \mid S \in \Sigma^{j-1}\} \qquad \text{for all } x \in X^{j-1}.$$

Then $f_{j-1}$ is a well-defined and continuous function on the compact set $X^{j-1}$ such that

$$\epsilon^j := \min\{f_{j-1}(x) \mid x \in X^{j-1}\} \quad \text{and}$$
$$X^j := \{x \in X^{j-1} \mid f_{j-1}(x) = \epsilon^j\}.$$

Analogous to the reasoning which was used in part (a), it follows from the induction hypothesis that $\epsilon^j$ is indeed well-defined and $X^j \neq \emptyset$, compact, convex. This completes the inductive proof of the lemma. □

LEMMA 7.4. Let $v \in G^n$ where $n \geq 2$.

(i) The critical number $\kappa$ is well-defined, i.e., $\kappa$ is finite.

(ii) The sets $\Sigma_j$ of coalitions are nonempty for $1 \leq j \leq \kappa$.

(iii) The real numbers $\epsilon^j$ form a decreasing sequence, i.e., $\epsilon^{j+1} < \epsilon^j$ for $1 \leq j \leq \kappa-1$.

PROOF. (a) We show that $\Sigma^{j-1} - \Sigma^j \neq \emptyset$ when $j \geq 1$, $\Sigma^j \neq \emptyset$. Let $\Sigma^j \neq \emptyset$ where $j \geq 1$. Write $\Sigma^j = \{S_1, S_2, \ldots, S_r\}$ where $r \geq 1$. Because $\Sigma^j = \Sigma^{j-1} - \Sigma_j$, we derive from (2.19) and (2.20) that for any $1 \leq k \leq r$ there exists $x^{(k)} \in X^j$ with $e(S_k, x^{(k)}) < \epsilon^j$. Then $e(S_k, x^{(i)}) \leq \epsilon^j$ for $i \neq k$. (2.23)

Define the vector $y \in \mathbb{R}^n$ by $y := r^{-1} \sum_{k=1}^{r} x^{(k)}$. Then $y \in X^j$ by the convexity of the set $X^j$ and hence, by (2.19), we can choose $S \in \Sigma^{j-1}$ with $e(S,y) = \epsilon^j$. Then also $e(S,x^{(k)}) \leq \epsilon^j$ for all $1 \leq k \leq r$. Now it follows that

$$\epsilon^j = v(S) - y(S) = r^{-1} \sum_{k=1}^{r} \left(v(S) - x^{(k)}(S)\right) \leq r^{-1} \sum_{k=1}^{r} \epsilon^j = \epsilon^j.$$

So, the last inequality must be an equality and therefore, $e(S,x^{(k)}) = \epsilon^j$ for all $1 \leq k \leq r$. Together with (2.23), this implies that $S \neq S_k$ for all $1 \leq k \leq r$. We conclude that $S \in \Sigma^{j-1} - \Sigma^j$.

(b) We have shown that $\Sigma^{j-1} - \Sigma^j \neq \emptyset$ when $j \geq 1$, $\Sigma^j \neq \emptyset$. Together with the inclusion $\Sigma^j \subset \Sigma^{j-1}$ for all $j \geq 1$, this establishes the finiteness of the critical number $\kappa$ because in each stage $j$ at least one coalition has to be removed from the set $\Sigma^{j-1}$ to obtain $\Sigma^j$. Moreover, the statement

(ii) holds because $\Sigma_j = \Sigma^{j-1} - \Sigma^j \neq \emptyset$ for all $1 \leq j \leq \kappa-1$, while $\Sigma_\kappa = \Sigma^{\kappa-1} - \Sigma^\kappa = \Sigma^{\kappa-1} \neq \emptyset$.

(c) Finally, we prove the validity of statement (iii). Let $1 \leq j \leq \kappa-1$. Then $\Sigma^j \neq \emptyset$ and so, we can apply part (a). By $y \in X^j$ and (2.18), we have $\epsilon^{j+1} \leq \max\{e(S_k,y) \mid 1 \leq k \leq r\}$. Further, (2.23) implies for all $1 \leq k \leq r$

$$e(S_k,y) = v(S_k) - y(S_k) = r^{-1} \sum_{i=1}^{r} \left(v(S_k) - x^{(i)}(S_k)\right) < \epsilon^j.$$

From this we conclude that $\epsilon^{j+1} < \epsilon^j$ whenever $1 \leq j \leq \kappa-1$.□

The finiteness of the procedure is based on the sequence $\Sigma^0 \supset \Sigma^1 \supset \Sigma^2 \supset \cdots \supset \Sigma^\kappa$ of strict inclusions between sets of coalitions, whereas the inclusions between sets of imputations in the sequence $X^0 \supset X^1 \supset X^2 \supset \cdots \supset X^\kappa$ are not necessarily strict as has been illustrated by the 3-person game of Example 4.4. The nonempty compact convex set $X^\kappa$ is all-important and is called the <u>lexicographic centre</u> of the game.

PROPOSITION 7.5. The lexicographic centre $X^\kappa$ of a game $v \in G^n$ where $n \geq 2$, is a singleton.

PROOF. Let $v \in G^n$ where $n \geq 2$. The set $\{S \subset N \mid S \neq N,\emptyset\}$ of nontrivial coalitions is partitioned into the sets $\Sigma_1, \Sigma_2, \cdots, \Sigma_\kappa$. Thus, for any $i \in N$ there exists a unique $1 \leq j_i \leq \kappa$ such that $\{i\} \in \Sigma_{j_i}$. By (2.20), this implies

$$v(\{i\}) - y_i = \epsilon^{j_i} \qquad \text{for all } i \in N \text{ and all } y \in X^{j_i}.$$

Let $x \in X^\kappa$. For all $i \in N$ we have $x \in X^\kappa \subset X^{j_i}$ and hence, $x_i = v(\{i\}) - \epsilon^{j_i}$. It follows that the set $X^\kappa$ contains at most one point. But $X^\kappa \neq \emptyset$ by Lemma 7.3(ii). We conclude that the set $X^\kappa$ is a singleton. □

With the aid of the next lemma, we establish at once that the lexicographic centre agrees with the nucleolus and as such, the nucleolus is always a singleton.

LEMMA 7.6.

(i) Let $1 \le j \le \kappa$, $x \in X^j$ and $y \in X^{j-1} - X^j$. Then

$$\theta(x) <_L \theta(y).$$

(ii) $\theta(x) <_L \theta(y)$ for all $x \in X^\kappa$ and all $y \in I(v) - X^\kappa$.

PROOF. (i) We distinguish the two cases $j = 1$ and $j \neq 1$.
Case one. Suppose $j = 1$. By (2.19) and $x \in X^1$, we have that $e(S,x) \le \epsilon^1$ for all $S \neq N,\emptyset$. Moreover, from $y \in X^0 - X^1$ we deduce the existence of a coalition $T \neq N,\emptyset$, such that $e(T,y) > \epsilon^1$. Further, $e(S,x) = e(S,y) = 0$ whenever $S = N$ or $S = \emptyset$. Now it follows immediately that $\theta(x) <_L \theta(y)$.

Case two. Suppose $2 \le j \le \kappa$. The set $\{S \subset N \mid S \neq N,\emptyset\}$ of nontrivial coalitions is partitioned into the sets $\Sigma_1, \cdots, \Sigma_{j-1}, \Sigma^{j-1}$. By (2.20) and Lemma 7.4(iii), we have

$$e(S,x) = e(S,y) = \epsilon^i \ge \epsilon^{j-1} \qquad \text{for all } S \in \Sigma_i,\ 1 \le i \le j-1.$$

Further, from (2.19) and Lemma 7.4(iii) we derive

$$e(S,x) \le \epsilon^j < \epsilon^{j-1} \quad \text{and} \quad e(S,y) \le \epsilon^{j-1} \qquad \text{for all } S \in \Sigma^{j-1}.$$

In addition, $y \in X^{j-1} - X^j$ implies the existence of a coalition $T \in \Sigma^{j-1}$ with $e(T,y) > \epsilon^j$. Now it follows that $\theta(x) <_L \theta(y)$. This completes the proof of statement (i).

(ii) Let $y \in I(v) - X^\kappa$. Then $y \in X^{j-1} - X^j$ for a certain $1 \le j \le \kappa$. Clearly, we have $X^\kappa \subset X^j$. By applying part (i), we obtain that $\theta(x) <_L \theta(y)$ for all $x \in X^\kappa$. □

THEOREM 7.7. The nucleolus of a game coincides with the lexicographic centre of the game. The nucleolus $\mathcal{N}(v)$ of a game v consists of a unique point which is denoted by $\eta(v)$.

The theorem is a direct consequence of Definition 7.1, Proposition 7.5 as well as Lemma 7.6(ii). The geometric characterization of the nucleolus indicates that the nucleolus is always included in the nonempty intersection of any strong $\epsilon$-core with the imputation set. A formal proof of this observation is given. In addition, we give an existence proof of the kernel and the bargaining set $\mathcal{M}$ which is provided by the fact that the nucleolus always belongs to the kernel.

THEOREM 7.8. Let $v \in G^n$ and $\epsilon \in \mathbb{R}$. Then

$$\eta(v) \in C_\epsilon(v) \cap I(v) \qquad \text{whenever } C_\epsilon(v) \cap I(v) \neq \emptyset.$$

In particular, $\eta(v) \in C(v)$ whenever $C(v) \neq \emptyset$.

PROOF. Let $C_\epsilon(v) \cap I(v) \neq \emptyset$. Choose $y \in C_\epsilon(v) \cap I(v)$ and put $x := \eta(v)$. Then $y \in I(v)$ and $e(S,y) \leq \epsilon$ for all $S \neq N,\emptyset$ by (2.7) and (1.6). The excesses of N and $\emptyset$ are of no importance in the lexicographic comparison between $\theta(x)$ and $\theta(y)$ because $e(S,x) = e(S,y) = 0$ whenever $S = N$ or $S = \emptyset$. So, we may ignore these two excesses at x and at y. By the definition of the complaint vectors, we have $e(S,x) \leq \theta_1(x)$ for all $S \neq N,\emptyset$, and $\theta_1(y) = e(T,y)$ for a certain $T \neq N,\emptyset$. Together with $\theta(x) \leq_L \theta(y)$, this implies for all $S \neq N,\emptyset$

$$e(S,x) \leq \theta_1(x) \leq \theta_1(y) = e(T,y) \leq \epsilon.$$

Hence, $x \in C_\epsilon(v) \cap I(v)$. In case $C(v) \neq \emptyset$, then we have $C(v) \cap I(v) = C(v) \neq \emptyset$ and thus, $x \in C(v)$. □

THEOREM 7.9. $\eta(v) \in \mathcal{K}(v)$ for all $v \in G^n$.

PROOF. Let $v \in G^n$ and put $x := \eta(v)$. Assume $x \notin \mathcal{K}(v)$. By (2.14), there exist $i,j \in N$, $i \neq j$, such that $x_i > v(\{i\})$ and $s_{ji}(x) > s_{ij}(x)$. Define the vector $y \in \mathbb{R}^n$ and the set $\Sigma$ of coalitions by

$$y := x - \alpha e^i + \alpha e^j \qquad \text{where}$$
$$\alpha := \min\left[x_i - v(\{i\}),\ \tfrac{1}{2}s_{ji}(x) - \tfrac{1}{2}s_{ij}(x)\right],$$
$$\Sigma := \{S \subset N \mid e(S,x) \geq s_{ji}(x) \text{ and } S \notin \Gamma_{ji}\}.$$

Then $\alpha > 0$, $y \neq x$ and $y \in I(v)$. Furthermore, we assert that

$$e(S,y) = e(S,x) \geq s_{ji}(x) \qquad \text{for all } S \in \Sigma, \tag{2.24}$$
$$e(S,y) < s_{ji}(x) \qquad \text{for all } S \notin \Sigma. \tag{2.25}$$

In order to prove (2.24) and (2.25), we first notice that

$$\begin{aligned} e(S,y) &= e(S,x) + \alpha && \text{if } S \in \Gamma_{ij} \\ &= e(S,x) - \alpha && \text{if } S \in \Gamma_{ji} \\ &= e(S,x) && \text{otherwise.} \end{aligned}$$

(i) To prove the statement (2.24), it is sufficient to show that $S \in \Sigma$ implies both $S \notin \Gamma_{ji}$ and $S \notin \Gamma_{ij}$. Let $S \in \Sigma$. Clearly, $S \notin \Gamma_{ji}$ by the definition of the set $\Sigma$, while $S \notin \Gamma_{ij}$ because of $e(S,x) \geq s_{ji}(x) > s_{ij}(x)$. Thus, (2.24) holds.

(ii) The statement (2.25) is explained by the facts that

$$\begin{aligned} e(S,y) &= e(S,x) - \alpha \leq s_{ji}(x) - \alpha < s_{ji}(x) && \text{if } S \in \Gamma_{ji} \\ &= e(S,x) + \alpha \leq s_{ij}(x) + \alpha < s_{ji}(x) && \text{if } S \in \Gamma_{ij} \\ &= e(S,x) < s_{ji}(x) \quad \text{if } S \notin \Sigma,\ S \notin \Gamma_{ji} && \text{and } S \notin \Gamma_{ij}. \end{aligned}$$

(iii) Choose $T \in \Gamma_{ji}$ such that $e(T,x) = s_{ji}(x)$. Then $T \notin \Sigma$ and hence,

$$\max\left[e(S,x) \mid S \notin \Sigma\right] = e(T,x) = s_{ji}(x). \qquad (2.26)$$

(iv) Put $k := |\Sigma|$. Now it follows from (2.24)-(2.26) that $\theta_i(y) = \theta_i(x)$ for $1 \leq i \leq k$ and $\theta_{k+1}(y) < s_{ji}(x) = \theta_{k+1}(x)$. We conclude that $\theta(y) <_L \theta(x)$ where $x = \eta(v)$ and $y \in I(v)$. However, this is in contradiction with the definition of the nucleolus. □

COROLLARY 7.10. $\mathcal{K}(v) \neq \emptyset$ and $\mathcal{M}(v) \neq \emptyset$ for all $v \in G^n$.

The Theorems 7.8 and 7.9 were already proved in Schmeidler (1969), while the above corollary is a direct consequence of the two Theorems 7.9 and 6.3. We also conclude that the nucleolus of the 3-person game v of Example 4.4 coincides with the unique kernel-element, i.e., $\eta(v) = \frac{1}{2}(2\alpha, 1-\alpha, 1-\alpha)$.

Furthermore, Theorem 7.9 together with Theorem 6.9 imply that the nucleolus possesses the substitution property and the dummy player property. The relative invariance under S-equivalence as well as the symmetry property are valid with respect to the nucleolus and can be shown in a straightforward way by using Definition 7.1. It is left to the reader to check these two properties of the nucleolus.

It is still an open problem to find an axiomatization of the nucleolus. A similar version of the nucleolus, called the prenucleolus, has been characterized in Sobolev (1975) by means of a so-called reduced game property for values on the class of all games.

The prenucleolus of a game $v \in G^n$ is defined as the set of all pre-imputations $x \in I^*(v)$ satisfying $\theta(x) \leq_L \theta(y)$ for all $y \in I^*(v)$. It is well-known that the prenucleolus of any game consists of a unique point. Obviously, the prenucleolus of a game coincides with the nucleolus whenever the prenucleolus is individually rational.
A reduced game property was also used in Peleg (1986) to axiomatize the prekernel on the class of all games. In fact the theory concerning the reduced games is of interest in itself, but it will not be considered in this work.

## 8. Balancedness

In the last section of this chapter we treat the notions of balanced collections and (totally) balanced games. The idea of balanced collections was introduced in Shapley (1967) who studied the relationship between the balanced collections of a game and the conditions that determine whether or not the game has a nonempty core. For any coalition $S \subset N$ we define the indicator function $1_S: N \to \{0,1\}$ by

$$1_S(i) := 1 \quad \text{if } i \in S$$
$$\qquad = 0 \quad \text{if } i \in N-S.$$

DEFINITION 8.1. Let $N = \{1,2,\ldots,n\}$ and $S \subset N$, $S \neq \emptyset$. A collection $\mathcal{C} = \{S_1,S_2,\ldots,S_m\}$ of distinct nonempty subsets of the coalition S is said to be balanced over S if there exist positive numbers $\alpha_1, \cdots, \alpha_m$, such that for all $i \in S$

$$\sum_{j;i\in S_j} \alpha_j = 1 \quad \text{or equivalently,} \quad \sum_{j=1}^{m} \alpha_j 1_{S_j}(i) = 1.$$

The associated positive numbers are called weights for the balanced collection $\mathcal{C}$. The weights are equal to one if and only if the balanced collection $\mathcal{C}$ over S is a partition of the set S. Therefore, balanced collections may be regarded as generalized partitions. For example, the collection $\{\{1,2\},\{1,3\},\{2,3\}\}$ is balanced over $\{1,2,3\}$ with unique weights which are equal to 0.5. However, there also exist balanced collections over $\{1,2,3\}$ which do not possess unique weights, e.g., $\{\{1\},\{2\},\{1,3\},\{2,3\}\}$.

DEFINITION 8.2. A game $v \in G^n$ is said to be <u>balanced</u> if for any balanced collection $\mathcal{C} = \{S_1, S_2, \ldots, S_m\}$ over the player set N with corresponding weights $\alpha_1, \alpha_2, \cdots, \alpha_m$, we have

$$\sum_{j=1}^{m} \alpha_j v(S_j) \leq v(N).$$

The balancedness condition for a game requires that it is not advantageous with respect to the earnings in the game to divide the grand coalition N into the sets of any balanced collection over N on the understanding that the earnings of the involved sets are weighted according to the corresponding weights.

We first show that the nonemptiness of the core is sufficient for the balancedness of the game.

THEOREM 8.3. A game $v \in G^n$ is balanced whenever $C(v) \neq \emptyset$.

PROOF. Let $v \in G^n$ be such that the core $C(v) \neq \emptyset$. Let $\mathcal{C} = \{S_1, S_2, \ldots, S_m\}$ be a balanced collection over the player set N with corresponding positive weights $\alpha_1, \alpha_2, \cdots, \alpha_m$. Choose $x \in C(v)$. Then $\sum_{j=1}^{m} \alpha_j 1_{S_j}(i) = 1$ for all $i \in N$, $x(N) = v(N)$ and $x(S_j) \geq v(S_j)$ for all $1 \leq j \leq m$. Now it follows that

$$\sum_{j=1}^{m} \alpha_j v(S_j) \leq \sum_{j=1}^{m} \alpha_j \sum_{i \in S_j} x_i = \sum_{j=1}^{m} \alpha_j \sum_{i \in N} x_i 1_{S_j}(i)$$

$$= \sum_{i \in N} x_i \sum_{j=1}^{m} \alpha_j 1_{S_j}(i) = \sum_{i \in N} x_i \cdot 1 = x(N) = v(N). \qquad \square$$

The nonemptiness of the core is a necessary condition for the balancedness of the game. The proof of this statement can be based on the duality theorem for linear programs.

THEOREM 8.4. If a game $v \in G^n$ is balanced, then $C(v) \neq \emptyset$.

PROOF. Let $v \in G^n$ be balanced. Put $m := 2^n - 1$ and write $2^N = \{\emptyset, S_1, S_2, \ldots, S_m\}$. Let the two vectors $b \in \mathbb{R}^n$, $c \in \mathbb{R}^m$ and the real-valued m×n matrix $A = [a_{ij}]_{i=1}^{m} {}_{j=1}^{n}$ be given by

$$b_j := 1, \qquad c_i := v(S_i), \qquad a_{ij} := 1_{S_i}(j)$$

for all $1 \le i \le m$, $1 \le j \le n$. Further, let the sets $G_p(A,b)$ and $G_d(A,c)$ be defined by

$$G_p(A,b) := \{y \in \mathbb{R}^m \mid yA = b \text{ and } y_i \ge 0 \text{ for all } 1 \le i \le m\},$$
$$G_d(A,c) := \{x \in \mathbb{R}^n \mid Ax \ge c\}.$$

Then we get

$$G_p(A,b) = \{y \in \mathbb{R}^m \mid \sum_{i=1}^{m} y_i 1_{S_i}(j) = 1 \text{ for all } 1 \le j \le n \text{ and also } y_i \ge 0 \text{ for all } 1 \le i \le m\},$$

$$G_d(A,c) = \{x \in \mathbb{R}^n \mid x(S) \ge v(S) \text{ for all } S \subset N,\ S \ne \emptyset\}.$$

Obviously, the set $G_d(A,c)$ is nonempty and in addition,

$$\min\Big(\sum_{j=1}^{n} b_j x_j \mid x \in G_d(A,c)\Big) = \min\big(x(N) \mid x \in G_d(A,c)\big) \ge v(N).$$

Due to this, the duality theorem for linear programs yields

$$\min\Big(\sum_{j=1}^{n} b_j x_j \mid x \in G_d(A,c)\Big) = \max\Big(\sum_{i=1}^{m} y_i c_i \mid y \in G_p(A,b)\Big).$$

Moreover, the positive coordinates of any $y \in G_p(A,b)$ can be regarded as the weights of the corresponding coalitions. So, any $y \in G_p(A,b)$ can be associated in a natural way with a balanced collection over N. Therefore, the balancedness of the game v implies

$$\max\Big(\sum_{i=1}^{m} y_i c_i \mid y \in G_p(A,b)\Big) = \max\Big(\sum_{i=1}^{m} y_i v(S_i) \mid y \in G_p(A,b)\Big)$$

$\le v(N)$. Now it follows that $\min\big(x(N) \mid x \in G_d(A,c)\big) = v(N)$. Hence, there exists a vector $x \in G_d(A,c)$ with $x(N) = v(N)$. As a consequence, $x \in C(v)$ and $C(v) \ne \emptyset$. □

COROLLARY 8.5. A game $v \in G^n$ is balanced if and only if $C(v) \ne \emptyset$.

The above relationship between the balancedness and the nonemptiness of the core is due to Bondareva (1963) as well as Shapley (1967). The class of balanced n-person games is denoted by $B^n$. Closely related to the balancedness is the notion of totally balancedness.

DEFINITION 8.6. A game $v \in G^n$ is said to be <u>totally balanced</u> if the induced subgames $(S;v_S)$ are balanced for all $S \subset N$, $S \neq \emptyset$.
Here the <u>subgame</u> $(S;v_S)$ with player set S is given by

$$v_S(T) := v(T) \qquad \text{for all } T \subset S.$$

In view of the above corollary, a game is totally balanced if and only if any subgame has a nonempty core. For instance, the 3-person game v of Example 4.4 is totally balanced. The class of totally balanced n-person games is characterized in Kalai and Zemel (1982a) as the class of flow games - games that are generated by flows in networks - and in Shapley and Shubik (1969) as the class of market games - games that are derived from exchange economies with money. Further, totally balanced games can be induced by

- certain assignment problems (Shapley and Shubik, 1972);
- certain linear production economies (Owen, 1975b);
- certain optimization problems (Kalai and Zemel, 1982b);
- certain controlled convex programming problems (Dubey and Shapley, 1984);
- certain sequencing and assignment problems (Tijs, Parthasarathy, Potters and Rajendra Prasad, 1984).

We conclude this section with a characterization of the nucleolus in terms of balanced collections which is due to Kohlberg (1971). For that purpose, we fix some notation with respect to the largest nontrivial excesses at a given imputation, the second largest nontrivial excesses, etc.

DEFINITION 8.7. Let $v \in G^n$ where $n \geq 2$. For any $x \in I(v)$ we define

$$\Sigma^0 := \{S \subset N \mid S \neq N, \emptyset\}, \qquad \mathcal{B}^0(x) := \emptyset,$$

$$\Sigma_0(x) := \{\{i\} \mid e^v(\{i\},x) = 0\}$$

and for $j = 1, 2, \cdots, \kappa(x)$, we define recursively

$$\epsilon^j(x) := \max\left[e^v(S,x) \mid S \in \Sigma^0 - \mathcal{B}^{j-1}(x)\right] \tag{2.27}$$

$$\Sigma_j(x) := \{S \in \Sigma^0 - \mathcal{B}^{j-1}(x) \mid e^v(S,x) = \epsilon^j(x)\} \tag{2.28}$$

$$\mathcal{B}^j(x) := \mathcal{B}^{j-1}(x) \cup \Sigma_j(x) = \bigcup_{i=1}^{j} \Sigma_i(x) \tag{2.29}$$

where $\kappa(x) := \min\left[j \mid j \geq 1,\ \mathcal{B}^j(x) = \Sigma^0\right].$ (2.30)

Clearly, the notions (2.27)-(2.30) are well-defined. The array $\Sigma_0(x), \Sigma_1(x), \cdots, \Sigma_{\kappa(x)}(x)$ of sets of coalitions is called the <u>coalition array</u> corresponding to the imputation x. According to the next theorem, certain properties of the coalition array determine whether or not the involved imputation is equal to the nucleolus of the game.

THEOREM 8.8. Let $v \in G^n$ where $n \geq 2$ and $x \in I(v)$. Then the following three statements are equivalent.

(i) $x = \eta(v)$.

(ii) For all $1 \leq j \leq \kappa(x)$ and all $y \in \mathbb{R}^n$ such that

$$y(N) = 0 \quad \text{and} \quad y(S) \geq 0 \quad \text{for all } S \in \Sigma_0(x) \cup \mathcal{B}^j(x),$$

we have $\quad y(S) = 0$ for all $S \in \mathcal{B}^j(x)$.

(iii) For all $1 \leq j \leq \kappa(x)$ there exists $\mathcal{B}_0^j(x) \subset \Sigma_0(x)$ such that $\mathcal{B}_0^j(x) \cup \mathcal{B}^j(x)$ is a balanced collection over N.

PROOF. We prove (iii) ⇒ (ii) ⇒ (i) ⇒ (iii).

(a) Suppose that (iii) holds. Let $1 \leq j \leq \kappa(x)$ and $y \in \mathbb{R}^n$ be such that

$$y(N) = 0 \quad \text{and} \quad y(S) \geq 0 \quad \text{for all } S \in \Sigma_0(x) \cup \mathcal{B}^j(x).$$

Let $T \in \mathcal{B}^j(x)$. We show $y(T) = 0$. In view of (iii), there exists a balanced collection $\mathcal{C}$ over N of the form

$$\mathcal{C} = \mathcal{B}_0^j(x) \cup \mathcal{B}^j(x) \qquad \text{where } \mathcal{B}_0^j(x) \subset \Sigma_0(x).$$

Denote the positive weights for the balanced collection $\mathcal{C}$ by $\alpha_S$, $S \in \mathcal{C}$. Then $\sum_{S\in\mathcal{C}} \alpha_S 1_S(i) = 1$ for all $i \in N$. Hence,

$$\sum_{S\in\mathcal{C}} \alpha_S y(S) = \sum_{S\in\mathcal{C}} \alpha_S \sum_{i\in S} y_i = \sum_{S\in\mathcal{C}} \alpha_S \sum_{i\in N} y_i 1_S(i)$$

$$= \sum_{i\in N} y_i \sum_{S\in\mathcal{C}} \alpha_S 1_S(i) = \sum_{i\in N} y_i \cdot 1 = y(N) = 0.$$

From this equality, $y(S) \geq 0$ and $\alpha_S > 0$ for all $S \in \mathcal{C}$, it follows that $y(S) = 0$ for all $S \in \mathcal{C}$. In particular, we get $y(T) = 0$. So, (iii) implies (ii).

(b) Suppose that (ii) holds. In order to prove $x = \eta(v)$, it suffices to show that

$$y \in I(v),\ \theta(y) \leq_L \theta(x) \qquad \text{imply} \qquad y = x.$$

Let $y \in I(v)$ such that $\theta(y) \leq_L \theta(x)$. Then $(y-x)(N) = 0$ and

$y_i - x_i \geq v(\{i\}) - x_i = e(\{i\},x) = 0$ for all $\{i\} \in \Sigma_0(x)$.

Further, we may ignore the two excesses of N and Ø at x and at y in the lexicographic comparison between $\theta(x)$ and $\theta(y)$ because these excesses are always zero. Now we assert that

$$(y-x)(S) = 0 \qquad \text{whenever } S \in \mathcal{B}^j(x),\ 1 \leq j \leq \kappa(x). \tag{2.31}$$

The assertion (2.31) is established by induction on j, $1 \leq j \leq \kappa(x)$. In case $S \in \mathcal{B}^1(x)$, then we have

$e(S,x) = \theta_1(x) \geq \theta_1(y) \geq e(S,y)$ and thus,

$(y-x)(S) = e(S,x) - e(S,y) \geq 0$ for all $S \in \mathcal{B}^1(x)$. Hence,

$(y-x)(N) = 0$ and $(y-x)(S) \geq 0$ for all $S \in \Sigma_0(x) \cup \mathcal{B}^1(x)$.

It follows from (ii) that $(y-x)(S) = 0$ for all $S \in \mathcal{B}^1(x)$. This completes the proof of the induction basis for $j = 1$. Let $2 \leq j \leq \kappa(x)$ and suppose that $(y-x)(T) = 0$ for all $T \in \mathcal{B}^{j-1}(x)$. Then $e(T,x) = e(T,y)$ for all $T \in \mathcal{B}^{j-1}(x)$. Together with $\theta(y) \leq_L \theta(x)$, this implies $e(S,x) \geq e(S,y)$ for all $S \in \Sigma_j(x)$. Hence,

$(y-x)(N) = 0$ and $(y-x)(S) \geq 0$ for all $S \in \Sigma_0(x) \cup \mathcal{B}^j(x)$.

It follows from (ii) that $(y-x)(S) = 0$ for all $S \in \mathcal{B}^j(x)$. This completes the inductive proof of the assertion (2.31). Because $\mathcal{B}^{\kappa(x)}(x) = \Sigma^0$ by (2.30), we conclude from (2.31) that $y(S) = x(S)$ for all $S \subset N$ and in particular, we get $y_i = x_i$ for all $i \in N$. Thus, $y = x$ whenever $y \in I(v)$ such that $\theta(y) \leq_L \theta(x)$. So, (ii) implies (i).

(c) The implication (i) ⇒ (iii) can be shown by using the theory of linear (in)equalities, but its proof is rather laborious and of no importance for the theory developed in this work. So, we omit the proof of the third implication (i) ⇒ (iii). □

EXAMPLE 8.9. We consider once again the 3-person game v of (2.10) in order to illustrate Theorem 8.8. We already know that the nucleolus of the game v is equal to the payoff vector $x = \frac{1}{2}(2\alpha, 1-\alpha, 1-\alpha)$. The excesses at x are given by

$$\begin{aligned} e(S,x) &= 0 && \text{if } S = N \text{ or } S = \emptyset \\ &= -\alpha && \text{if } S = \{1\} \\ &= \alpha-1 && \text{if } S = \{2,3\} \\ &= \tfrac{1}{2}(\alpha-1) && \text{otherwise.} \end{aligned}$$

In case $0.5 < \alpha < 1$, then it follows immediately from the formulas (2.27)-(2.30) that

$$\begin{aligned} & & \Sigma_0(x) &= \emptyset, \\ \epsilon^1(x) &= \tfrac{1}{2}(\alpha-1), & \Sigma_1(x) &= \{\{2\},\{3\},\{1,2\},\{1,3\}\}, \\ \epsilon^2(x) &= \alpha-1, & \Sigma_2(x) &= \{\{2,3\}\}, \\ \epsilon^3(x) &= -\alpha, & \Sigma_3(x) &= \{\{1\}\} \quad \text{and} \quad \kappa(x) = 3. \end{aligned}$$

According to part (iii) of Theorem 8.8, the three relevant collections $\Sigma_1(x)$, $\Sigma_1(x) \cup \Sigma_2(x)$ and $\Sigma_1(x) \cup \Sigma_2(x) \cup \Sigma_3(x)$ are balanced over $\{1,2,3\}$. It is straightforward to check the balancedness of the three collections. Notice that the numbers $\epsilon^j(x)$ and the sets $\Sigma_j(x)$, $1 \le j \le 3$, are precisely the numbers $\epsilon^j$ and the sets $\Sigma_j$, $1 \le j \le 3$, corresponding to the geometric determination of the nucleolus with the aid of the procedure described by the formulas (2.18)-(2.22). In fact, the real numbers $\epsilon^j(x)$ and the sets $\Sigma_j(x)$ of Definition 8.7 agree with the real numbers $\epsilon^j$ and the sets $\Sigma_j$ of Definition 7.2 whenever the imputation x is equal to the nucleolus of the game.

THEOREM 8.10. Let $v \in G^n$ where $n \ge 2$ and $x = \eta(v)$. Then

$$\kappa(x) = \kappa \quad \text{and} \quad \epsilon^j(x) = \epsilon^j,\ \Sigma_j(x) = \Sigma_j \quad \text{for all } 1 \le j \le \kappa.$$

PROOF. The statement is shown by induction on j, $1 \le j \le \kappa$. For all $y, z \in I(v)$ we have $e(S,y) = e(S,z) = 0$ whenever $S = N$ or $S = \emptyset$ and hence, we may ignore these two excesses at y and at z in the lexicographic comparison between the complaint vectors $\theta(y)$ and $\theta(z)$.
(i) First of all, we prove the induction basis for $j = 1$. For any $y \in I(v)$ we have $\theta(x) \le_L \theta(y)$ and therefore,

$$\epsilon^1(x) = \max_{S\in\Sigma^0} e(S,x) = \theta_1(x) \le \theta_1(y) = \max_{S\in\Sigma^0} e(S,y).$$

It follows that $\epsilon^1 = \min_{y \in I(v)} \max_{S \in \Sigma^0} e(S,y) \geq \epsilon^1(x)$. Further, $\epsilon^1 \leq \epsilon^1(x)$ because $x \in I(v)$. We conclude that $\epsilon^1 = \epsilon^1(x)$. Together with (2.19) and (2.20), this implies

$$X^1 = \{y \in I(v) \mid \max_{S \in \Sigma^0} e(S,y) = \epsilon^1(x)\} \qquad \text{and}$$

$$\Sigma_1 = \{S \in \Sigma^0 \mid e(S,y) = \epsilon^1(x) \text{ for all } y \in X^1\}.$$

Then $x \in X^1$ and so, $\Sigma_1 \subset \{S \in \Sigma^0 \mid e(S,x) = \epsilon^1(x)\} = \Sigma_1(x)$. Now we show that $\Sigma_1 = \Sigma_1(x)$. Assume $\Sigma_1 \neq \Sigma_1(x)$. Choose $T \in \Sigma_1(x) - \Sigma_1$. Put $\Sigma^0 - \Sigma_1 = \{S_1, S_2, \ldots, S_r\}$ where $S_1 = T$. We derive from (2.19) and (2.20) that for any $1 \leq k \leq r$ there exists $y^{(k)} \in X^1$ such that

$$e(S_k, y^{(k)}) < \epsilon^1. \text{ Then } e(S_k, y^{(i)}) \leq \epsilon^1 \text{ for } i \neq k.$$

Define the vector $y \in \mathbb{R}^n$ by $y := r^{-1} \sum_{k=1}^{r} y^{(k)}$. In view of Lemma 7.3(ii), the set $X^1$ is convex and therefore, $y \in X^1$. Furthermore, we have that for all $1 \leq k \leq r$

$$e(S_k, y) = v(S_k) - y(S_k) = r^{-1} \sum_{i=1}^{r} \left[v(S_k) - y^{(i)}(S_k)\right] < \epsilon^1.$$

Because $\epsilon^1 = \epsilon^1(x)$, it follows that $y \in X^1$ satisfies

$$e(S,y) = \epsilon^1(x) = e(S,x) \qquad \text{if } S \in \Sigma_1,$$

$$e(S,y) < \epsilon^1(x) = e(T,x) \qquad \text{if } S \in \Sigma^0 - \Sigma_1.$$

We conclude that $\theta(y) <_L \theta(x)$ where $x = \eta(v)$ and $y \in I(v)$. However, this is in contradiction with the definition of the nucleolus. We have shown the equality $\Sigma_1 = \Sigma_1(x)$ which completes the proof of the induction basis for $j = 1$.

(ii) Let $2 \leq j \leq \kappa$ and suppose that $\epsilon^i(x) = \epsilon^i$ as well as $\Sigma_i(x) = \Sigma_i$ for all $1 \leq i \leq j-1$. We prove $\epsilon^j(x) = \epsilon^j$ and $\Sigma_j(x) = \Sigma_j$. In view of (2.21) and (2.29), we have

$$\Sigma^{j-1} = \Sigma^0 - \bigcup_{i=1}^{j-1} \Sigma_i = \Sigma^0 - \bigcup_{i=1}^{j-1} \Sigma_i(x) = \Sigma^0 - \mathcal{B}^{j-1}(x). \tag{2.32}$$

By both (2.32) and (2.27), we obtain $\epsilon^j(x) = \max_{S\in\Sigma^{j-1}} e(S,x)$.

By Theorem 7.7, $x \in X^\kappa$ and thus, $x \in X^i$ for all $0 \le i \le \kappa$. For any $y \in X^{j-1}$ and any $S \in \Sigma_i$, $1 \le i \le j-1$, it follows from (2.20) and (2.28) that $e(S,y) = \epsilon^i = \epsilon^i(x) = e(S,x)$. Together with $\theta(x) \le_L \theta(y)$ for all $y \in I(v)$, this implies

$$\max_{S\in\Sigma^{j-1}} e(S,x) \le \max_{S\in\Sigma^{j-1}} e(S,y) \qquad \text{for all } y \in X^{j-1}.$$

Hence, $\epsilon^j = \min_{y\in X^{j-1}} \max_{S\in\Sigma^{j-1}} e(S,y) \ge \max_{S\in\Sigma^{j-1}} e(S,x) = \epsilon^j(x)$.

Further, $\epsilon^j \le \epsilon^j(x)$ because of $x \in X^{j-1}$. We conclude that $\epsilon^j = \epsilon^j(x)$. From this, $x \in X^j$, (2.20), (2.28) and (2.32), we deduce

$$\begin{aligned}\Sigma_j &= \{S \in \Sigma^{j-1} \mid e(S,y) = \epsilon^j(x) \text{ for all } y \in X^j\} \\ &\subset \{S \in \Sigma^{j-1} \mid e(S,x) = \epsilon^j(x)\} = \Sigma_j(x).\end{aligned}$$

In order to establish that $\Sigma_j = \Sigma_j(x)$, assume $\Sigma_j \ne \Sigma_j(x)$. Analogous to the reasoning which was used in part (i), we can construct $y \in X^j$ and $T \in \Sigma_j(x) - \Sigma_j$ such that

$$\begin{aligned} e(S,y) &= \epsilon^i(x) = e(S,x) && \text{if } S \in \Sigma_i,\ 1 \le i \le j-1, \\ e(S,y) &= \epsilon^j(x) = e(S,x) && \text{if } S \in \Sigma_j, \\ e(S,y) &< \epsilon^j(x) = e(T,x) && \text{if } S \in \Sigma^{j-1} - \Sigma_j. \end{aligned}$$

From this we conclude that $\theta(y) <_L \theta(x)$. Since this is in contradiction with $x = \eta(v)$, we have shown $\Sigma_j = \Sigma_j(x)$. This completes the inductive proof.

(iii) Similar to (2.32), we have $\Sigma^j = \Sigma^0 - \mathcal{B}^j(x)$ for all $1 \le j \le \kappa$. Now it follows from (2.22) and (2.30) that the equality $\kappa = \kappa(x)$ holds. □

Finally, we remark that the core on the class of all (totally) balanced games has been axiomatized in Peleg (1985, 1986) by means of so-called reduced game properties. A third version of balancedness, called quasibalancedness, is treated in Section III.2. The idea of quasibalancedness forms the basis of an additional one-point solution concept for cooperative games in characteristic function form. The next chapter is devoted to the study of that significant value.

# CHAPTER III

# THE $\tau$-VALUE

## 1. The upper vector, the concession vector and the gap function of a game

In the previous chapter we have reviewed various well-known solution concepts for cooperative games. The one-point or set-valued solution concepts, discussed in the second chapter, were being introduced during the period 1944-1972. This chapter is devoted to another one-point solution concept which was introduced in Tijs (1981). The development of the theory concerning the so-called $\tau$-value is due to Driessen and Tijs. The $\tau$-value is based on the ideas of an upper bound for the core and the excesses with respect to this upper bound. The involved upper bound for the core of an n-person game is given in terms of the worth of the player set itself and the (n-1)-person coalitions.

DEFINITION 1.1. Let $v \in G^n$. The <u>upper vector</u> $b^v \in \mathbb{R}^n$ and the <u>gap function</u> $g^v: 2^N \to \mathbb{R}$ of the game v are given by

$$b_i^v := v(N) - v(N-\{i\}) \qquad \text{for all } i \in N, \quad (3.1)$$

$$g^v(S) := \sum_{j \in S} b_j^v - v(S) = -e^v(S,b^v) \quad \text{for all } S \subset N. \quad (3.2)$$

The i-th coordinate $b_i^v$ of the upper vector $b^v$ is called the <u>marginal contribution of player</u> i (with respect to the grand coalition N) in the game v. The term upper vector is explained by the fact that the vector $b^v$ is an upper bound for the core of the game v, as stated in the next lemma.

LEMMA 1.2. Let $v \in G^n$.
(i) If $x \in C(v)$, then $x_i \le b_i^v$ for all $i \in N$.
(ii) $C(v) = \emptyset$ when there exists $S \subset N$ such that $g^v(S) < 0$.
(iii) $g^v(N-\{i\}) = g^v(N)$ for all $i \in N$.

PROOF. (i) Let $x \in C(v)$ and $i \in N$. By (3.1) and (2.8), we have

$$b_i^v = v(N) - v(N-\{i\}) = x(N) - v(N-\{i\}) \geq x_i .$$

(ii) Let $S \subset N$ be such that $g^v(S) < 0$. We prove $C(v) = \emptyset$. Assume $C(v) \neq \emptyset$. Choose $x \in C(v)$. Then $x(S) \geq v(S)$ by (2.8), while part (i) implies $x(S) \leq b^v(S)$. Now it follows that

$$b^v(S) \geq v(S) \qquad \text{or equivalently,} \qquad g^v(S) \geq 0.$$

This is in contradiction with $g^v(S) < 0$. Hence, $C(v) = \emptyset$.
(iii) We deduce from the two formulas (3.2) and (3.1) that

$$g^v(N-\{i\}) = b^v(N-\{i\}) - v(N-\{i\})$$
$$= b^v(N) - v(N) = g^v(N) \qquad \text{for all } i \in N.$$ □

The expression $g^v(S)$ is called the <u>gap of coalition</u> S in the game v. Note that the simple equality $g^v(\emptyset) = 0$ holds. A nonnegative (nonpositive respectively) gap of S in the game v represents the loss (gain) to the coalition S if its members withdraw from the upper payoff vector $b^v$ in order to form their own coalition. Thus, although the game itself refers to savings, the corresponding gap function refers to losses with respect to the upper vector of the game.

In view of Lemma 1.2(ii), the condition that the gap function is nonnegative is necessary for the nonemptiness of the core. This condition is also sufficient whenever $n = 2$, but in case $n > 2$, then this condition is in general not sufficient for the game to possess a nonempty core.

Recall that the maximal excesses at any imputation are used to define the nucleolus. Here the maximal excesses at the upper vector are directly related to the minimal gaps. Given any player we pay special attention to the minimal gaps among the gaps of coalitions containing the player. It appears that the relevant minimal gaps induce a lower bound for the core.

DEFINITION 1.3. Let $v \in G^n$. The <u>concession vector</u> $\lambda^v \in \mathbb{R}^n$ of the game v is given by

$$\lambda_i^v := \min\{g^v(S) \mid S \subset N,\ i \in S\} \qquad \text{for all } i \in N. \quad (3.3)$$

LEMMA 1.4. $b_i^v - \lambda_i^v \leq x_i$ for all $v \in G^n$, $i \in N$ and $x \in C(v)$.

PROOF. Let $v \in G^n$, $i \in N$ and $x \in C(v)$. By the definition of the concession vector, there exists a coalition $S \subset N$ such that $i \in S$ and $g^v(S) = \lambda_i^v$. Since $x \in C(v)$, we have

$$\lambda_i^v = g^v(S) = b^v(S) - v(S) \geq (b^v - x)(S) \geq b_i^v - x_i$$

where the last inequality follows from Lemma 1.2(i) and by noting that $i \in S$. □

According to Lemma 1.4, the vector $b^v - \lambda^v$ is a lower bound for the core of the game v because the resulting payoff by any core-element to any player i is at least the amount $b_i^v - \lambda_i^v$. Thus, for any balanced game v, the expression $\lambda_i^v$ can be regarded as the maximal concession made by player i with respect to his upper payoff. The induced lower bound $b^v - \lambda^v$ is called the <u>disagreement</u> <u>vector</u> of the game v. We conclude this section with two useful properties of the upper vector, the gap function and the concession vector. The proofs of these two properties are straightforward and are left to the reader.

LEMMA 1.5. Let $v \in G^n$, $w \in G^n$, $\alpha \in \mathbb{R}$ and $d \in \mathbb{R}^n$. Then

(i) $b^{v+w} = b^v + b^w$, $g^{v+w} = g^v + g^w$, and $\lambda^{v+w} \geq \lambda^v + \lambda^w$.

(ii) $b^{\alpha v+d} = \alpha b^v + d$, $g^{\alpha v+d} = \alpha g^v$.

If $\alpha \geq 0$, then $\lambda^{\alpha v+d} = \alpha\lambda^v$.

By Lemma 1.5, both the upper vector and the gap function behave in a linear way, i.e., the maps $v \to b^v$ and $v \to g^v$ on $G^n$ are linear. We often write b, $\lambda$, g instead of $b^v$, $\lambda^v$, $g^v$ whenever it is clear which game v is meant.

## 2. The $\tau$-value of a quasibalanced game

If a game v is superadditive, then we obtain that $b_i^v = v(N) - v(N-\{i\}) \geq v(\{i\})$ or equivalently, $g^v(\{i\}) \geq 0$ for all $i \in N$ and hence, any player i prefers his marginal contribution to the amount $v(\{i\})$ what player i can attain for himself. However, many superadditive games v satisfy $b^v(N) > v(N)$ or equivalently, $g^v(N) > 0$ which strict inequality implies that it is not possible to distribute the amount v(N) among the players in such a way that each player gets at least his marginal contribution. Thus, in case a game is superadditive such that the gap of the grand coalition is positive, then the upper vector is preferred by the players, but it is not an efficient payoff vector. We call the upper vector $b^v$ an <u>utopia payoff vector</u> for the game v whenever $g^v(N) > 0$ and $g^v(\{i\}) \geq 0$ for all $i \in N$. With respect to this utopia payoff vector $b^v$, the grand coalition N has to make a concession the size of $g^v(N)$ $(= b^v(N) - v(N))$. Now we suppose that the contribution of any single player to the joint concession amount $g^v(N)$ will be determined by minimization of the gaps of coalitions containing the single player. This supposition can be explained as follows.

"Player i promises the other members of any coalition S containing player i their utopia payoffs whenever they cooperate with him. Player i will keep the remaining part of the associated worth v(S) and hence, player i himself gets the amount

$$v(S) - \sum_{j \in S-\{i\}} b_j^v \qquad \text{or equivalently,} \qquad b_i^v - g^v(S).$$

The maximal amount what player i can achieve in this way is obtained whenever the formed coalition has a minimal gap among the gaps of coalitions containing player i."

By the above reasoning and (3.3), any player i can achieve the payoff $b_i^v - \lambda_i^v$. Therefore, the maximal concession of player i with respect to his utopia payoff is equal to the amount $\lambda_i^v$. We require that the concession amounts $\lambda_i^v$, $i \in N$, are nonnegative, i.e., $g^v(S) \geq 0$ for all $S \subset N$. We also require that these maximal concession amounts total at least as much as the joint concession amount, i.e., $\lambda^v(N) \geq g^v(N)$. The two conditions give rise to introduce the notion of quasibalancedness.

DEFINITION 2.1. The class $QB^n$ of <u>quasibalanced n-person games</u> is given by

$$QB^n := \{v \in G^n \mid \lambda^v(N) \geq g^v(N) \text{ and } g^v(S) \geq 0 \text{ for all } S \subset N\} \tag{3.4}$$

$$= \{v \in G^n \mid \lambda_i^v \geq 0 \text{ for all } i \in N \text{ and } (b^v - \lambda^v)(N) \leq v(N) \leq b^v(N)\}. \tag{3.5}$$

For any quasibalanced game v, the expression $\lambda_i^v$ of (3.3) can be interpreted as the maximal contribution of player i to the joint concession amount $g^v(N)$. It suffices to charge each player only a fraction of his maximal contribution because of the inequality $\lambda^v(N) \geq g^v(N)$. The $\tau$-value of the game occurs if this fraction is the same for all players. Formally, the $\tau$-value of a quasibalanced game is defined as follows.

DEFINITION 2.2. The <u>$\tau$-value</u> $\tau(v) \in \mathbb{R}^n$ of a game $v \in QB^n$ is given by

$$\tau(v) := b^v \qquad \text{if } g^v(N) = 0$$

$$= b^v - g^v(N)\Big(\sum_{j\in N} \lambda_j^v\Big)^{-1}\lambda^v \qquad \text{if } g^v(N) > 0. \quad (3.6)$$

The $\tau$-value of a quasibalanced game was introduced in Tijs (1981). However, Tijs did not use the gap function in the original definition of both the quasibalancedness and the $\tau$-value. With the original characterization (3.5) of quasibalancedness in mind, Tijs defined the $\tau$-value $\tau(v)$ of a quasibalanced game v as the unique pre-imputation lying on the straight line segment with end points the upper vector $b^v$ and the disagreement vector $b^v - \lambda^v$ of the game v (see Figure 3.1).

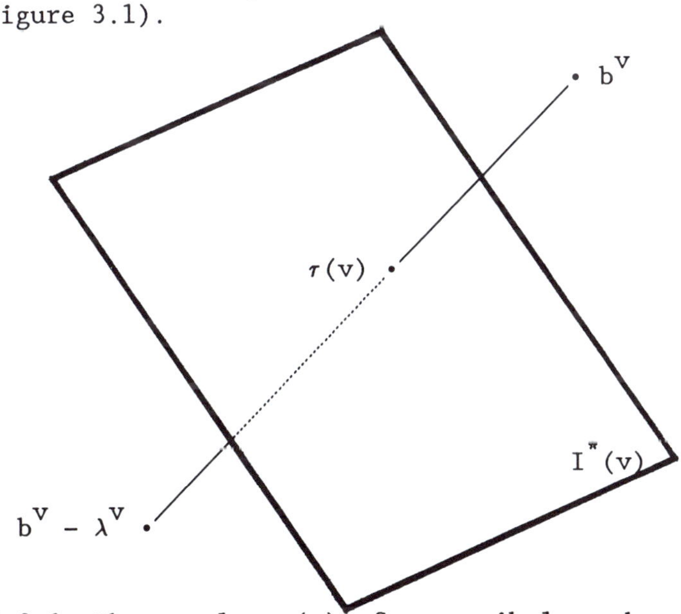


FIGURE 3.1. The $\tau$-value $\tau(v)$ of a quasibalanced game v.

In case the game possesses a nonempty core, then the upper (disagreement respectively) vector should be regarded as an upper (lower) bound for the core of the game and therefore, the $\tau$-value represents some efficient compromise between two bounds for the core. Evidently, the formal definition of the $\tau$-value agrees with this geometric characterization of the $\tau$-value. Furthermore, by this geometric viewpoint, the notion of the $\tau$-value resembles the solution of Kalai and Smorodinsky (1975) for two-person bargaining problems.

### 3. Necessary and sufficient conditions for the $\tau$-value on $QB^n$ to belong to the core

In this section we investigate the conditions that determine whether or not the $\tau$-value of a quasibalanced game is included in the core. First of all, we establish that balancedness implies quasibalancedness.

PROPOSITION 3.1. $B^n \subset QB^n$.

PROOF. Let $v \in B^n$. Then $C(v) \neq \emptyset$ by Corollary II.8.5. Choose $x \in C(v)$. By Lemma 1.4, we get that $b_i - \lambda_i \leq x_i$ for all $i \in N$. Thus, $\lambda(N) \geq b(N) - x(N) = b(N) - v(N) = g(N)$. Further, it follows from Lemma 1.2(ii) that $g(S) \geq 0$ for all $S \subset N$. Due to the quasibalancedness condition (3.4), we conclude that $v \in QB^n$. □

Before we determine the dimension of the classes of (quasi) balanced n-person games, we look at the structure of these subclasses of $G^n$. Generally speaking, a collection G of n-person games is called a <u>cone</u> in the linear space $G^n$ if

$v \in G$, $w \in G$ and $\alpha \in [0,\infty)$ imply $v + w \in G$ and $\alpha v \in G$.

In addition, a collection G of n-person games is said to be a <u>polyhedral</u> <u>cone</u> in $G^n$ if G can be described as the intersection of a finite number of weak inequalities that are linear in the worths of the coalitions for all games in G.

PROPOSITION 3.2. $B^n$ and $QB^n$ are polyhedral cones in $G^n$.

PROOF. (i) Because the number of balanced collections over the player set N is finite, it follows immediately from Definition II.8.2 that the class $B^n$ is a polyhedral cone in the linear space $G^n$.

(ii) Due to (3.4), $v \in QB^n$ iff $g(S) \geq 0$ for all $S \subset N$ and $\lambda(N) \geq g(N)$. From the formulas (3.1) and (3.2) we derive that

$$g(S) = |S|v(N) - v(S) - \sum_{j\in S} v(N-\{j\}) \quad \text{for all } S \subset N.$$

By using (3.3), the inequality $\lambda(N) \geq g(N)$ is equivalent to

$$\sum_{j\in N} g(S_j) \geq g(N) \quad \text{for all n-tuples } (S_1,S_2,\ldots,S_n) \text{ of}$$
coalitions satisfying $i \in S_i \subset N$ for all $i \in N$.

In a straightforward way, it follows that $v \in QB^n$ if and only if

$$v(S) + \sum_{j\in S} v(N-\{j\}) \leq |S|v(N) \quad \text{for all } S \subset N \quad \text{and}$$

$$\sum_{j\in N} \Big(v(S_j) + \sum_{k\in S_j-\{j\}} v(N-\{k\})\Big) \leq \Big(1-n + \sum_{j\in N} |S_j|\Big)v(N)$$

for all n-tuples $(S_1,S_2,\ldots,S_n)$ of coalitions satisfying $i \in S_i \subset N$ for all $i \in N$.

From this we directly conclude that the class $QB^n$ is a polyhedral cone in the linear space $G^n$. □

Further, we remark that the cone structure of the class $B^n$ ($QB^n$ respectively) can also be established with the aid of Corollary II.8.5 (Lemma 1.5 and (3.4)). The two classes $B^n$ and $QB^n$ are full-dimensional cones in the linear space $G^n$.

THEOREM 3.3. $\dim B^n = \dim QB^n = 2^n - 1$.

PROOF. We consider the unanimity n-person games of (1.5) and the unit vectors in $\mathbb{R}^n$ as well. Let $T \subset N$, $T \neq \emptyset$. For any player $i \in T$ we have that the corresponding i-th unit vector $e^i \in C(u_T)$ because

$$e^i(S) = 1 = u_T(S) \quad \text{if } S \supset T,$$
$$e^i(S) \geq 0 = u_T(S) \quad \text{otherwise.}$$

Thus, $C(u_T) \neq \emptyset$. Now it follows from Corollary II.8.5 and Proposition 3.1 that $u_T \in B^n$ and $u_T \in QB^n$ for all $T \neq \emptyset$. Recall that the set $\{u_T \in G^n \mid T \subset N, T \neq \emptyset\}$ forms a basis of the $(2^n-1)$-dimensional linear space $G^n$. We conclude that $B^n$ as well as $QB^n$ are full-dimensional cones in $G^n$. □

The nonnegativity of the gap function is a necessary condition, but in general not a sufficient condition for the nonemptiness of the core. In case the game possesses a nonnegative gap function such that the gap of the grand coalition is equal to zero, then the core of the game is a singleton consisting of the $\tau$-value.

PROPOSITION 3.4. Let $v \in G^n$ be such that

$$g^v(S) \geq 0 \text{ for all } S \subset N \qquad \text{and} \qquad g^v(N) = 0.$$

Then $C(v) = \{\tau(v)\} = \{b^v\}$.

PROOF. By (3.2), $b(N) = v(N)$ and $b(S) \geq v(S)$ for all $S \subset N$. So, $b \in C(v)$ because of (2.8). By Lemma 1.2(i), we have that $x_i \leq b_i$ for all $i \in N$ and all $x \in C(v)$. Together with the equality $x(N) = v(N) = b(N)$, this implies the vector equality $x = b$ when $x \in C(v)$. We conclude that $C(v) = \{b\}$. Therefore, $v \in B^n \subset QB^n$, while $\tau(v) = b$ since $g(N) = 0$. □

For quasibalanced games of which the gap of the player set is positive, we state a necessary and sufficient condition for the $\tau$-value to belong to the core. The condition in question is expressed in terms of the gap function and the concession vector of the game.

THEOREM 3.5. Let $v \in QB^n$ be such that $g^v(N) > 0$. Then

$\tau(v) \in C(v)$ iff

$[g^v(N)]^{-1}\lambda^v(N) \geq [g^v(S)]^{-1}\lambda^v(S)$ for all $S \subset N$

satisfying $g^v(S) > 0$ and $2 \leq |S| \leq n-2$.

PROOF. Put $x := \tau(v)$. By the formula (3.6) for the $\tau$-value, we have $x(N) = v(N)$ and $b_i - \lambda_i \leq x_i \leq b_i$ for all $i \in N$. Together with the formulas (3.1)-(3.3), this yields that

$$x(N-\{i\}) = v(N) - x_i \geq v(N) - b_i = v(N-\{i\}) \text{ as well as}$$

$$x_i \geq b_i - \lambda_i \geq b_i - g(\{i\}) = v(\{i\}) \qquad \text{for all } i \in N.$$

From this we conclude that $x \in C(v)$ if and only if

$$x(S) \geq v(S) \qquad \text{for all } S \subset N \text{ with } 2 \leq |S| \leq n-2.$$

In view of (3.6) and (3.2), we obtain that $x \in C(v)$ iff

$$g(S) \geq g(N)[\lambda(N)]^{-1}\lambda(S) \quad \text{whenever } 2 \leq |S| \leq n-2. \quad (3.7)$$

From (3.3) we derive that $g(S) = 0$ implies $\lambda_i = 0$ for all $i \in S$ and in particular, $\lambda(S) = 0$ whenever $g(S) = 0$. So, the equality holds in (3.7) for all $S \subset N$ with $g(S) = 0$. Now the equivalence of the two statements in the theorem follows immediately from (3.7). □

COROLLARY 3.6. Let $n \in \{1,2,3\}$. Then

$$\tau(v) \in C(v) \qquad \text{for all } v \in QB^n.$$

The corollary is a direct consequence of Proposition 3.4 and Theorem 3.5. We also observe that any quasibalanced game with at most three players is balanced. Together with the known inclusion $B^n \subset QB^n$ for all $n \in \mathbb{N}$, this yields the equality $B^n = QB^n$ whenever $n \leq 3$. In other words, the quasibalancedness condition for an n-person game where $n \leq 3$ is equivalent to the balancedness condition for the game. In case $n \geq 4$, then the inclusion $B^n \subset QB^n$ is strict because there exist quasibalanced n-person games with an empty core.

EXAMPLE 3.7. Let the 4-person game v be given by

$$v(12) = v(13) = v(23) = v(123) = v(N) = 3,$$

$$v(S) = 0 \text{ otherwise.}$$

Then $b = (3,3,3,0)$, $g(4) = 0$, $g(S) = 6$ whenever $|S| \geq 3$, and $g(S) = 3$ for all other $S \neq \emptyset$. By (3.3), we obtain $\lambda = (3,3,3,0) = b$ and hence, $\lambda(N) = 9 \geq 6 = g(N)$. Now it follows from (3.4) and (3.6) respectively that $v \in QB^4$ and $\tau(v) = \frac{1}{3}b = (1,1,1,0)$. It is left to the reader to verify that the core of the game v is empty, i.e., $C(v) = \emptyset$.

The next example even illustrates that the $\tau$-value of a balanced game v does not necessarily belong to the core of the game, although the $\tau$-value is some efficient compromise between the upper bound $b^v$ and the lower bound $b^v - \lambda^v$ for the core of the game v. As a matter of fact, two different games with identical cores may possess different $\tau$-values, one of them lying outside the core.

EXAMPLE 3.8. Let the 4-person game v be given by

$$v(12) = v(13) = v(23) = v(123) = 2,$$

$$v(N) = 3 \text{ and } v(S) = 0 \text{ otherwise.}$$

Because the game v is not superadditive, we also consider the associated superadditive 4-person game w given by

$$w(124) = w(134) = w(234) = 2, \quad w(S) = v(S) \text{ otherwise.}$$

It is straightforward to verify that the cores of both games coincide, i.e., $C(v) = C(w) = \{(1,1,1,0)\}$. In order to determine the $\tau$-values of the balanced games v and w, we notice that $b^v = (3,3,3,1)$ and $b^w = (1,1,1,1)$, while the corresponding gap functions $g^v$ and $g^w$ are as follows.

| coalition S | 1 | 2 | 3 | 4 | 12 | 13 | 14 | 23 | 24 | 34 | $\lvert S\rvert \geq 3$ |
|---|---|---|---|---|---|---|---|---|---|---|---|
| gap $g^v(S)$ | 3 | 3 | 3 | 1 | 4 | 4 | 4 | 4 | 4 | 4 | 7 |
| gap $g^w(S)$ | 1 | 1 | 1 | 1 | 0 | 0 | 2 | 0 | 2 | 2 | 1 |

By using the two formulas (3.3) and (3.6), we obtain

$$\lambda^v = (3,3,3,1) = b^v, \qquad \lambda^w = (0,0,0,1),$$

$$b^v - \lambda^v = (0,0,0,0), \qquad b^w - \lambda^w = (1,1,1,0),$$

$$\tau(v) = 0.3b^v = 0.3(3,3,3,1), \qquad \tau(w) = (1,1,1,0).$$

Thus, $\tau(v) \neq \tau(w)$ although $C(v) = C(w)$. Moreover, we have $\tau(w) \in C(w)$ and $\tau(v) \notin C(v)$. Note that it is not surprising that the $\tau$-value of the game v lies outside the core since the two bounds $b^v$ and $b^v - \lambda^v$ are not sharp bounds for the core which is a singleton. Generally speaking, the $\tau$-value may fall outside the core although the upper vector and the disagreement vector are sharp bounds for the core, while the core itself is large.

Finally, we show that the $\tau$-value $\tau(v)$ does not belong to the bargaining set $\mathcal{M}(v)$. For that purpose, we prove that $x_4 = 0$ whenever $x \in \mathcal{M}(v)$. Let $x \in I(v)$ be such that $x_4 > 0$. Then $x(123) < 3$. So, $x(12) < 2$ or $x(13) < 2$ or $x(23) < 2$. Say $x(12) < 2$ which is equivalent to $e^v(12,x) > 0$. From the positivity of the involved excess $e^v(12,x)$ we conclude that player 1 can object against player 4 with respect to the imputation x by using the coalition $\{1,2\}$. However, there exists no counterobjection to this objection because

$$e^v(T,x) = v(T) - x(T) = -x(T) \leq -x_4 < 0$$

$$\text{for all } T \subset N \text{ satisfying } 4 \in T,\ 1 \notin T.$$

Hence, $x \notin \mathcal{M}(v)$ whenever $x \in I(v)$ such that $x_4 > 0$. From this, it follows that $\tau(v) \notin \mathcal{M}(v)$ and further, $\tau(v) \notin \mathcal{K}(v)$ in view of Theorem II.6.3.

Another way to check whether or not the $\tau$-value lies inside the core is to look at the so-called fixed points of a suitably chosen map.

DEFINITION 3.9. The <u>maximum excess map</u> $ME^v\colon \mathbb{R}^n \to \mathbb{R}^n$ of a game $v \in G^n$ is as follows: for all $x \in \mathbb{R}^n$ and all $i \in N$

$$\begin{aligned} ME_i^v(x) &:= x_i + \max\{e^v(S,x) \mid S \subset N,\ i \in S\} \\ &= \max\{v(S) - x(S-\{i\}) \mid S \subset N,\ i \in S\}. \end{aligned} \tag{3.8}$$

The i-th coordinate $ME_i^v(x)$ of the vector $ME^v(x)$ represents the largest possible payoff to player i by withdrawing from the payoff vector x and forming a coalition containing player i, on the understanding that the other members of the formed coalition are satisfied with the amount they had according to the original payoff vector x. The next three properties of the maximum excess map follow immediately from its definition.

LEMMA 3.10. Let $v \in G^n$ and $x \in \mathbb{R}^n$.

(i) Let $i \in N$. Then $ME_i^v(x) \geq x_i$ iff there exists $S \subset N$ such that $i \in S$ and $e^v(S,x) \geq 0$.

(ii) $e^v(N,x) \geq 0$ implies $ME_i^v(x) \geq x_i$ for all $i \in N$.

(iii) $ME_i^v(x) \leq x_i$ for all $i \in N$ iff $e^v(S,x) \leq 0$ for all $S \subset N$.

PROPOSITION 3.11. Let $v \in G^n$ and $x \in \mathbb{R}^n$. Then

$$x \in C(v) \quad \text{iff} \quad e^v(N,x) = 0 \text{ and } ME^v(x) = x.$$

PROOF. By (2.8) and (1.6), we have that $x \in C(v)$ iff

$$e(N,x) = 0 \quad \text{and} \quad e(S,x) \leq 0 \text{ for all } S \subset N.$$

However, in case $e(N,x) = 0$, then the additional statement $e(S,x) \leq 0$ for all $S \subset N$ is equivalent to the equality $ME(x) = x$ because of Lemma 3.10(ii) and (iii). □

A point $x \in \mathbb{R}^n$ which satisfies the equality $ME^v(x) = x$, is called a <u>fixed point</u> of the map $ME^v$. From the above proposition we conclude that the core of a game consists of efficient fixed points of the corresponding maximum excess map. In particular, the $\tau$-value of a quasibalanced game belongs to the core of the game if and only if the $\tau$-value is a fixed point of the corresponding maximum excess map.

COROLLARY 3.12. Let $v \in QB^n$. Then

$$\tau(v) \in C(v) \qquad \text{iff} \qquad ME^v\big(\tau(v)\big) = \tau(v).$$

In view of Proposition 3.11, the core of a game is empty if and only if there exist no efficient fixed points of the corresponding maximum excess map. Nevertheless, the maximum excess map of any (not necessarily balanced) game always possesses fixed points.

THEOREM 3.13. $\{x \in \mathbb{R}^n \mid ME^v(x) = x\} \neq \emptyset$ for all $v \in G^n$.

PROOF. Let $v \in G^n$. Define the real number $\alpha$ by

$$\alpha := \max\big(|S|^{-1}v(S) \mid S \subset N,\ S \neq \emptyset\big)$$

and let the vector $x^{(0)} \in \mathbb{R}^n$ be given by $x_i^{(0)} := \alpha$ for all $i \in N$. For $j = 1, 2, \cdots, r$, we define recursively

$$\mathcal{W}_j := \{S \subset N \mid e(S,x^{(j-1)}) = 0\}$$

$$\mathcal{N}_j := \{i \in N \mid i \notin S \text{ for all } S \in \mathcal{W}_j\}$$

and so long as $\mathcal{N}_j \neq \emptyset$

$$\rho^j := \min\big(-|S \cap \mathcal{N}_j|^{-1}e(S,x^{(j-1)}) \mid S \notin \mathcal{W}_j,\ S \cap \mathcal{N}_j \neq \emptyset\big)$$

$x^{(j)} \in \mathbb{R}^n$ such that

$$x_i^{(j)} := x_i^{(j-1)} - \rho^j \qquad \text{if } i \in \mathcal{N}_j$$

$$\qquad = x_i^{(j-1)} \qquad \text{if } i \in N-\mathcal{N}_j$$

where $r := \min\big(j \mid j \geq 1,\ \mathcal{N}_j = \emptyset\big)$.

(i) We first show by induction on $j$, $j \geq 0$, that

$$e(S,x^{(j)}) \leq 0 \qquad \text{for all } S \subset N \text{ and all } j \geq 0. \tag{3.9}$$

We have that $e(S,x^{(0)}) = v(S) - x^{(0)}(S) = v(S) - |S|\alpha \leq 0$ for all $S \subset N$ where the inequality follows directly from the definition of the real number $\alpha$. This completes the proof of the induction basis for $j = 0$.

Let $j \geq 1$ and suppose that $e(S,x^{(j-1)}) \leq 0$ for all $S \subset N$.

By using the definition of the vector $x^{(j)}$, we obtain that $e(S,x^{(j)}) = v(S) - x^{(j)}(S) = v(S) - x^{(j-1)}(S) + |S \cap \mathcal{N}_j|\rho^j$ $= e(S,x^{(j-1)}) + |S \cap \mathcal{N}_j|\rho^j$ for $S \subset N$. If $S \cap \mathcal{N}_j = \emptyset$, then $e(S,x^{(j)}) = e(S,x^{(j-1)}) \leq 0$ due to the induction hypothesis. If $S \cap \mathcal{N}_j \neq \emptyset$, then $S \notin \mathcal{W}_j$ by the definition of the set $\mathcal{N}_j$ and hence, $e(S,x^{(j)}) = e(S,x^{(j-1)}) + |S \cap \mathcal{N}_j|\rho^j \leq 0$ where the inequality is due to the definition of the number $\rho^j$. This completes the inductive proof of the statement (3.9).
(ii) In order to prove the correctness of the procedure, we show that the critical number r is well-defined, i.e., r is finite. Clearly, $r = 1$ if $\mathcal{N}_1 = \emptyset$. In case $\mathcal{N}_1 \neq \emptyset$, we assert

$$\mathcal{N}_j \neq \emptyset \qquad \text{implies} \qquad \mathcal{N}_{j+1} \subset \mathcal{N}_j \text{ and } \mathcal{N}_{j+1} \neq \mathcal{N}_j. \tag{3.10}$$

To prove (3.10), suppose $\mathcal{N}_j \neq \emptyset$. If $S \in \mathcal{W}_j$, then $S \cap \mathcal{N}_j = \emptyset$ and consequently,

$$e(S,x^{(j)}) = e(S,x^{(j-1)}) + |S \cap \mathcal{N}_j|\rho^j = e(S,x^{(j-1)}) = 0.$$

So, $S \in \mathcal{W}_j$ implies $S \in \mathcal{W}_{j+1}$. Thus, $\mathcal{W}_j \subset \mathcal{W}_{j+1}$ and from this, it follows immediately that the inclusion $\mathcal{N}_{j+1} \subset \mathcal{N}_j$ holds. Next we show $\mathcal{N}_{j+1} \neq \mathcal{N}_j$. Note that $\{i\} \notin \mathcal{W}_j$ for all $i \in \mathcal{N}_j$ and so, the real number $\rho^j$ is well-defined. Choose $T \subset N$ such that $T \notin \mathcal{W}_j$, $T \cap \mathcal{N}_j \neq \emptyset$ and $e(T,x^{(j-1)}) = -|T \cap \mathcal{N}_j|\rho^j$. It follows that $e(T,x^{(j)}) = e(T,x^{(j-1)}) + |T \cap \mathcal{N}_j|\rho^j = 0$. Therefore, we obtain $T \in \mathcal{W}_{j+1}$. From this we conclude that $i \notin \mathcal{N}_{j+1}$ for all $i \in T \cap \mathcal{N}_j$ and hence, $\mathcal{N}_{j+1} \neq \mathcal{N}_j$. We have shown that (3.10) holds. Now the finiteness of the critical number r is a direct consequence of the strict inclusions as described by the statement (3.10).
(iii) Let $i \in N$. Because $\mathcal{N}_r = \emptyset$, there exists $T \in \mathcal{W}_r$ such that $i \in T$. That is, there exists $T \subset N$ such that $i \in T$ and $e(T,x^{(r-1)}) = 0$. Furthermore, $e(S,x^{(r-1)}) \leq 0$ for all $S \subset N$ because of (3.9). Together with Lemma 3.10(i) and (iii), this yields the equality $ME_i^v(x^{(r-1)}) = x_i^{(r-1)}$. We conclude that $x^{(r-1)}$ is a fixed point of the involved map $ME^v$. □

We consider once again the 4-person games of the Examples 3.7 and 3.8 in order to illustrate Proposition 3.11 and Corollary 3.12. By straightforward calculations, we obtain that the maximum excess map of the 4-person game of Example 3.7 has no efficient fixed points, although the set of all fixed points equals

$$\{(\alpha,\alpha,3-\alpha,0),\ (\alpha,3-\alpha,\alpha,0),\ (3-\alpha,\alpha,\alpha,0) \mid \tfrac{3}{2} \le \alpha \le 3\}.$$

As a consequence, the core of the game is empty. For the 4-person games v and w of Example 3.8 we deduce in a straightforward way that the two sets of all fixed points of the corresponding maximum excess maps coincide and are equal to

$$\{(\alpha,\alpha,2-\alpha,0),\ (\alpha,2-\alpha,\alpha,0),\ (2-\alpha,\alpha,\alpha,0) \mid 1 \le \alpha \le 2\}.$$

We recall that $\tau(v) = 0.3(3,3,3,1)$ and $\tau(w) = (1,1,1,0)$. We observe that the $\tau$-value of the game w coincides with the unique efficient fixed point $(1,1,1,0)$ and therefore, $\tau(w) \in C(w)$. However, the $\tau$-value of the game v is not a fixed point because of $\tau_4(v) \ne 0$ and thus, $\tau(v) \notin C(v)$.

## 4. An axiomatic characterization of the $\tau$-value on $QB^n$

In this section we present an axiomatization of the $\tau$-value on $QB^n$. For that purpose, we first show that the $\tau$-value, seen as a value on the class $QB^n$ of quasibalanced n-person games, possesses the standard properties (i)-(iv) for values as mentioned in Section II.1. Moreover, we look at the substitution property of the $\tau$-value on $QB^n$.

THEOREM 4.1. The $\tau$-value $\tau\colon QB^n \to \mathbb{R}^n$ possesses the next five properties: (i) individual rationality; (ii) symmetry; (iii) dummy player property; (iv) relative invariance under S-equivalence and (v) substitution property.

PROOF. Let $v \in QB^n$. Then $b_i^v - \lambda_i^v \le \tau_i(v) \le b_i^v$ for all $i \in N$ by the formula (3.6) for the $\tau$-value.

(i) Let $i \in N$. By the formulas (3.3) and (3.2), we have

$\lambda_i^v \le g^v(\{i\}) = b_i^v - v(\{i\})$. Hence, we deduce

$$\tau_i(v) \ge b_i^v - \lambda_i^v \ge v(\{i\}).$$

So, the $\tau$-value is individually rational.

(ii) Let $\theta$: N → N be a permutation. By straightforward calculations using the formulas (3.1)-(3.3), we obtain that

$$b^{\theta v}_{\theta(i)} = b^v_i \qquad \text{for all } i \in N,$$

$$g^{\theta v}(\theta S) = g^v(S) \qquad \text{for all } S \subset N,$$

$$\lambda^{\theta v}_{\theta(i)} = \lambda^v_i \qquad \text{for all } i \in N.$$

From this and (3.4) we conclude that $\theta v \in QB^n$ iff $v \in QB^n$. Further, it follows immediately from the formula (3.6) that

$$\tau_{\theta(i)}(\theta v) = \tau_i(v) \qquad \text{for all } i \in N.$$

So, the $\tau$-value is symmetric.

(iii) Let $i \in N$ be a dummy player in the game v. Then in particular, $v(N) - v(N-\{i\}) = v(\{i\})$. Now it follows that

$\tau_i(v) \le b^v_i = v(N) - v(N-\{i\}) = v(\{i\})$, while

$\tau_i(v) \ge v(\{i\})$ by applying part (i). Hence,

$\tau_i(v) = v(\{i\})$ for a dummy player i in the game v.

So, the $\tau$-value possesses the dummy player property.

(iv) Let $\alpha \in (0,\infty)$ and $d \in \mathbb{R}^n$. By Lemma 1.5(ii), we have $b^{\alpha v+d} = \alpha b^v + d$, $g^{\alpha v+d} = \alpha g^v$, as well as $\lambda^{\alpha v+d} = \alpha\lambda^v$. From this and (3.4) we conclude that $\alpha v + d \in QB^n$ iff $v \in QB^n$. Further, $\tau(\alpha v + d) = \alpha\tau(v) + d$ by using the formula (3.6). The $\tau$-value is relatively invariant under S-equivalence.

(v) Let $i \in N$ and $j \in N$ be substitutes in the game v. From the formulas (2.17), (3.1) and (3.2), we derive

$$b^v_i = v(N) - v(N-\{i\}) = v(N) - v(N-\{j\}) = b^v_j,$$

$$g^v(S \cup \{i\}) = b^v(S \cup \{i\}) - v(S \cup \{i\})$$

$$= b^v(S \cup \{j\}) - v(S \cup \{j\}) = g^v(S \cup \{j\})$$

for all $S \subset N-\{i,j\}$. From this and (3.3), it follows that $\lambda^v_i = \lambda^v_j$. In view of the two equalities $b^v_i = b^v_j$, $\lambda^v_i = \lambda^v_j$, and the formula (3.6), it is evident that $\tau_i(v) = \tau_j(v)$. So, the $\tau$-value possesses the substitution property. □

We recall that the lower bound $b^v - \lambda^v$ for the core of the balanced game v of Example 3.8 vanishes because $\lambda^v = b^v$ and as a consequence, the $\tau$-value $\tau(v)$ of the game v is proportional to the upper vector $b^v$. It is obvious from the formula (3.6) that the $\tau$-value of a quasibalanced game is proportional to the upper vector of the game whenever the maximal contribution of any player to the joint concession amount is precisely his upper payoff. According to the next theorem, this property as well as the relative invariance under S-equivalence fully characterize the $\tau$-value on the class of quasibalanced n-person games.

THEOREM 4.2. The $\tau$-value $\tau$: $QB^n \to \mathbb{R}^n$ is the unique value $\psi$ on $QB^n$ with the following two properties:
- relative invariance under S-equivalence and
- for any $v \in QB^n$ satisfying $\lambda^v = b^v$, the value vector $\psi(v)$ is proportional to the upper vector $b^v$.

PROOF. It remains to prove the uniqueness part. Suppose that $\psi$: $QB^n \to \mathbb{R}^n$ is a value with the two mentioned properties. We show that $\psi = \tau$ on $QB^n$. Let $v \in QB^n$. Define the vector $d \in \mathbb{R}^n$ and the game $w \in G^n$ by $d := b^v - \lambda^v$ and $w := v - d$. Because both $\psi$ and $\tau$ are relatively invariant under S-equivalence, we obtain $\psi(w) = \psi(v) - d$ as well as $\tau(w) = \tau(v) - d$. In order to prove $\psi(v) = \tau(v)$, it is sufficient to show $\psi(w) = \tau(w)$. By Lemma 1.5(ii), we have

$$b^w = b^v - d, \quad g^w = g^v, \quad \lambda^w = \lambda^v. \qquad \text{It follows that}$$

$$\lambda^w = \lambda^v = b^v - d = b^w \qquad \text{and in addition,}$$

$$v \in QB^n \text{ implies } w \in QB^n.$$

Since $w \in QB^n$ satisfies $\lambda^w = b^w$, the two value vectors $\psi(w)$ and $\tau(w)$ are proportional to the upper vector $b^w$. That is, there exist real numbers $\alpha$ and $\beta$ such that

$$\psi(w) = \alpha b^w \qquad \text{and} \qquad \tau(w) = \beta b^w.$$

Notice that $b_i^w = \lambda_i^w \geq 0$ for all $i \in N$. If $b^w(N) = 0$, then $b_i^w = 0$ for all $i \in N$ and hence, we obtain $\psi(w) = \tau(w)$. In case $b^w(N) > 0$, the efficiency property for $\psi$ and $\tau$ yields $\alpha b^w(N) = w(N) = \beta b^w(N)$ and thus, $\alpha = \beta$. Hence, we obtain $\psi(w) = \tau(w)$ whenever $b^w(N) > 0$. We have shown $\psi(w) = \tau(w)$ which equality completes the proof. □

Analogous to Theorem 4.2, the $\tau$-value on the class $QB^n$ can also be characterized with the aid of the property that the value is proportional to the concession vector whenever the upper vector vanishes. The corresponding proof is similar to the proof of Theorem 4.2 applied to $d = b^v$ instead of

$d = b^v - \lambda^v$ for all $v \in QB^n$. So, the proof of the second axiomatization of the $\tau$-value on the class of quasibalanced n-person games is left to the reader.

THEOREM 4.3. The $\tau$-value $\tau$: $QB^n \to \mathbb{R}^n$ is the unique value $\psi$ on $QB^n$ with the following two properties:
- relative invariance under S-equivalence and
- for any $v \in QB^n$ satisfying $b_i^v = 0$ for all $i \in N$, the value vector $\psi(v)$ is proportional to the vector $\lambda^v$.

## 5. 1-Convex games

The definition of the concession vector of a game in terms of the corresponding gap function gives rise to look at the class of n-person games for which the concession vector is characterized merely by the grand coalition, i.e., the gap of the grand coalition N is minimal among the gaps of nonempty coalitions. As usual, we also require that the gap function is nonnegative.

DEFINITION 5.1. The class $C_1^n$ of <u>1-convex</u> <u>n-person</u> <u>games</u> is given by (3.11)

$$C_1^n := \{v \in G^n | \ 0 \le g^v(N) \le g^v(S) \text{ for all } S \subset N, \ S \ne \emptyset\}.$$

The 1-convexity condition (3.11) for a game v is defined in terms of the corresponding gap function $g^v$, but it is also equivalent to the requirements $b^v(N) \ge v(N)$ and

$$v(S) \le v(N) - b^v(N-S) \qquad \text{for all } S \subset N, \ S \ne \emptyset. \quad (3.12)$$

The condition (3.12) states that the remaining part of the total savings for any nonempty coalition S in the game v is at least its worth v(S) whenever the total savings v(N) of the grand coalition are distributed in such a way that all players outside S receive their payoffs according to the upper vector $b^v$ of the game. From the condition (3.12) we deduce that the coalition constraints for the core of a 1-convex n-person game can be reduced to the (n-1)-person coalition constraints.

PROPOSITION 5.2. Let $v \in C_1^n$ and $x \in I^*(v)$. Then

$$x \in C(v) \qquad \text{iff} \qquad x_i \le b_i^v \text{ for all } i \in N.$$

PROOF. Suppose that $x_i \le b_i$ for all $i \in N$. From this and the condition (3.12), it follows that for all $S \subset N$, $S \neq \emptyset$,

$$v(S) \le v(N) - b(N-S) \le v(N) - x(N-S) = x(S).$$

So, $x \in C(v)$ by (2.8). Furthermore, the converse statement is always valid because of Lemma 1.2(i). □

THEOREM 5.3. Let $v \in C_1^n$ and $1_n := (1,1,\dots,1) \in \mathbb{R}^n$. Then

$$v \in QB^n \qquad \text{and} \qquad \tau(v) = b^v - n^{-1}g^v(N)1_n \in C(v).$$

PROOF. From (3.11) and (3.3) we derive that $\lambda_i = g(N)$ for all $i \in N$. Then we get $\lambda(N) = ng(N) \ge g(N)$ since $g(N) \ge 0$. Now it follows from (3.4) and (3.6) respectively that both $v \in QB^n$ and $\tau(v) = b - n^{-1}g(N)1_n$. By Proposition 5.2 and the inequality $g(N) \ge 0$, it is clear that $\tau(v) \in C(v)$.□

Concerning the $\tau$-value of a 1-convex game v, the players contribute equally to the joint concession amount of the grand coalition. This result is due to the fact that the maximal concession $\lambda_i^v$ of any player i with respect to his upper payoff $b_i^v$ is the same for all players. In particular, the core of a 1-convex game is nonempty because the $\tau$-value is included in the core. In point of fact, the notion of 1-convexity can be characterized in terms of the core and suitably chosen efficient payoff vectors.

THEOREM 5.4. $v \in C_1^n$ iff $b^v - g^v(N)e^i \in C(v)$ for all $i \in N$.

PROOF. Let $v \in G^n$. For any $i \in N$ we put $x^{(i)} := b - g(N)e^i$. Suppose that $x^{(i)} \in C(v)$ for all $i \in N$. We show that (3.11) holds. Let $S \subset N$, $S \neq \emptyset$. Then there exists $j \in S$. Now it follows from $x^{(j)} \in C(v)$ that $v(S) \le x^{(j)}(S) = b(S) - g(N)$ which is equivalent to $g(N) \le g(S)$. Further, we deduce from $x^{(j)} \in C(v)$ and Lemma 1.2(i) that $b_j \ge x_j^{(j)} = b_j - g(N)$ or equivalently, $g(N) \ge 0$. So, the condition (3.11) holds. We conclude that $v \in C_1^n$ whenever $x^{(i)} \in C(v)$ for all $i \in N$. Moreover, the converse statement is a direct consequence of Proposition 5.2. □

By Theorem 5.4, a necessary and sufficient condition for the 1-convexity of an n-person game v is that the vectors $b^v - g^v(N)e^i$, $i \in N$, belong to the core of the game. Note that each vector can be obtained from the upper vector by letting merely *one* coordinate decrease in such a way that an efficient payoff vector is arrived at. In view of the very last observation, the prefix *1-* is used. The remaining term convexity will be explained in Section V.1. A further development of the theory concerning the 1-convex games will be given in Chapter VII. Finally, we illustrate the notion of 1-convexity by two examples of 1-convex games.

EXAMPLE 5.5. Consider the n-person game v of (1.1) where $|P| \le \alpha(|Q| - 1)$. By the formulas (1.1), (3.1) and (3.2), we get

$$\begin{array}{llll} v(N) & = |P|, & & \\ v(N-\{i\}) & = |P| - 1 & \text{and} \quad b_i = 1 & \text{for all } i \in P, \\ v(N-\{i\}) & = |P| & \text{and} \quad b_i = 0 & \text{for all } i \in Q, \\ g(N) = b(N) - v(N) & = |P| - |P| = 0 & & \text{and further,} \\ g(S) = b(S) - v(S) & = |S \cap P| - v(S) \ge 0 & & \text{for all } S \subset N \end{array}$$

where the inequality follows from the definition of the game. Thus, the corresponding gap function is nonnegative such that $g(N) = 0$. So, the game v is 1-convex by (3.11), while $C(v) = \{\tau(v)\} = \{b\}$ by Proposition 3.4.

EXAMPLE 5.6. Let the 3-person game v be given by

$$v(i) = 0 \quad \text{for all } i \in N, \qquad v(N) = 18,$$
$$v(12) = v(13) = 9 \quad \text{and} \quad v(23) = 15.$$

Then $b = (3,9,9)$, $g(2) = g(3) = 9$, and $g(S) = 3$ for all other $S \neq \emptyset$. In view of (3.11), the game v is 1-convex and hence, $\tau(v) = (2,8,8) \in C(v)$ by Theorem 5.3. By (2.8), we get

$$C(v) = \{x \in \mathbb{R}^3_+ \mid x_1 + x_2 + x_3 = 18,\ x_1 \le 3,\ x_2 \le 9,\ x_3 \le 9\}$$
$$= \text{conv}\,\{(0,9,9),\ (3,6,9),\ (3,9,6)\}.$$

The core of the game v is drawn in Figure 3.2. Note that the core is a triangle inside the imputation set and the $\tau$-value coincides with the centre of gravity of the three extreme points of the core.

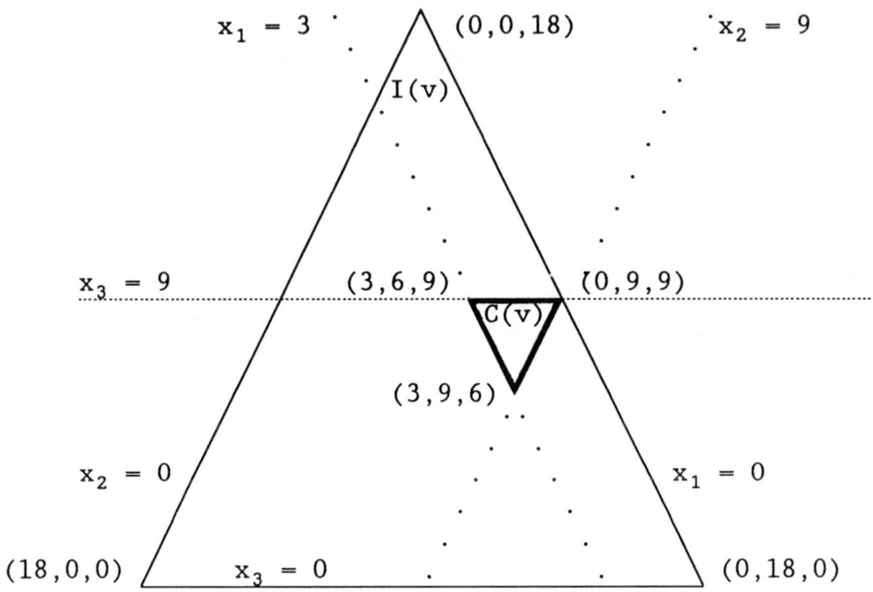


FIGURE 3.2. The indicated triangle represents the core $C(v)$ of the 1-convex 3-person game $v$ of Example 5.6.

## 6. Semiconvex games

The definition of the concession vector of a game in terms of the corresponding gap function inspires us to pay attention to the class of n-person games for which any coordinate of the concession vector is determined by the relevant one-person coalition, i.e., the gap of any single player is minimal among the gaps of coalitions containing the involved player. As usual, the gap function is also required to be nonnegative.

DEFINITION 6.1. The class $SC^n$ of <u>semiconvex n-person games</u> is given by

$$SC^n := \{v \in G^n \mid 0 \le g^v(\{i\}) \le g^v(S) \quad \text{for all } i \in N \text{ and all } S \subset N \text{ such that } i \in S\}. \tag{3.13}$$

The semiconvexity condition (3.13) for a game $v$ is equivalent to the requirements

$$b_i^v \ge v(\{i\}) \quad \text{and} \quad v(S) - b^v(S-\{i\}) \le v(\{i\})$$
$$\text{for all } i \in N \text{ and all } S \subset N \text{ with } i \in S. \tag{3.14}$$

The condition (3.14) expresses that any player in the game prefers his upper payoff to his single worth and further, the remaining part of the total savings for any member of a coalition is at most his minimal payoff (i.e., his single worth) whenever the other members of the coalition receive their maximal payoffs (i.e., their upper payoffs) for the formation of the coalition. The interrelationship of the semiconvexity to the term convexity will be discussed in Section V.1. The next theorem deals with the $\tau$-value on the class of semiconvex games.

THEOREM 6.2. Let $v \in SC^n$ satisfy $I(v) \neq \emptyset$ and $g^v(N) > 0$. Then

(i) $v \in QB^n$.

(ii) $\tau(v) = b^v - g^v(N)\left[\sum_{j\in N} g^v(\{j\})\right]^{-1}\left[g^v(\{1\}),\ldots,g^v(\{n\})\right]$.

(iii) $\tau(v) = v(N)\left[b^v(N)\right]^{-1}b^v$ whenever $v(\{i\}) = 0$ for all $i \in N$.

(iv) In addition, let $v(\{i\}) = 0$ for all $i \in N$. Then $\tau(v) \in C(v)$ iff $\left[g^v(N)\right]^{-1}b^v(N) \geq \left[g^v(S)\right]^{-1}b^v(S)$ for all $S \subset N$ satisfying $g^v(S) > 0$ and $2 \leq |S| \leq n-2$.

PROOF. By both (3.3) and (3.13), $\lambda_i = g(\{i\})$ for all $i \in N$.

(a) To prove the quasibalancedness condition (3.4), it is sufficient to show $\lambda(N) \geq g(N)$. We have

$$\lambda(N) = \sum_{j\in N} g(\{j\}) = b(N) - \sum_{j\in N} v(\{j\}) \geq b(N) - v(N) = g(N)$$

where the inequality follows from $I(v) \neq \emptyset$. Hence, $v \in QB^n$. So, (i) holds. The statement (ii) is a direct consequence of the formula (3.6) and $\lambda_i = g(\{i\})$ for all $i \in N$.

(b) To prove (iii) and (iv), let $v(\{i\}) = 0$ for all $i \in N$. Then we get that $g(\{i\}) = b_i - v(\{i\}) = b_i$ for all $i \in N$. Now it follows from part (ii) that the $\tau$-value is given by

$$\tau(v) = b - g(N)\left[b(N)\right]^{-1}b = v(N)\left[b(N)\right]^{-1}b.$$

So, (iii) holds. Further, part (iv) is a direct consequence of Theorem 3.5 and $\lambda_i = g(\{i\}) = b_i$ for all $i \in N$. □

In view of Theorem 6.2(iii), the $\tau$-value of a zero-normalized semiconvex game is proportional to the upper vector of the game. Note that the balanced 4-person game v of Example 3.8 is zero-normalized as well as semiconvex, but its $\tau$-value does not belong to the core. According to the next theorem, this is due to the fact that the game v is not zero-monotonic.

THEOREM 6.3. $\tau(v) \in C(v)$ if $v \in SC^4$ is zero-monotonic.

PROOF. Let $v \in SC^4$ be zero-monotonic. We define the vector $d \in \mathbb{R}^4$ and the 4-person game w by $d_i := v(\{i\})$ for all $i \in N$ and $w := v - d$. By Lemma 1.5(ii), we have $g^w = g^v$. Together with $v \in SC^4$, this implies $w \in SC^4$. Furthermore, $w(\{i\}) = 0$ for all $i \in N$ and by the zero-monotonicity of the game v, we obtain $0 \le w(S) \le w(T)$ whenever $S \subset T \subset N$. In particular, $w(N) \ge 0$ and hence, $I(w) \ne \emptyset$.
We show $\tau(w) \in C(w)$ by using Theorem 6.2. If $g^w(N) = 0$, then $\tau(w) \in C(w)$ because of Proposition 3.4. It remains to consider the case $g^w(N) > 0$. In view of Theorem 6.2(iv), it suffices for $\tau(w) \in C(w)$ to prove that

$$g^w(S)b^w(N) \ge g^w(N)b^w(S) \quad \text{for all } S \subset N \text{ with } |S| = 2.$$

Let $S \subset N$ be such that $|S| = 2$. Then we have

$$b_i^w = g^w(\{i\}) \ge 0 \quad \text{for all } i \in N,$$

$$b^w(S) = \sum_{j\in S} g^w(\{j\}) \le 2g^w(S) = 2b^w(S) - 2w(S),$$

$$b^w(N-S) = 2w(N) - \sum_{j\in N-S} w(N-\{j\}) \le 2w(N) - 2w(S)$$

where the last inequalities follow from the semiconvexity and the monotonicity of the game w respectively. Now we conclude that

$$b^w(S) \ge 2w(S) \ge 0 \quad \text{as well as}$$

$$2\big(w(N) - w(S)\big) \ge b^w(N-S) \ge 0 \quad \text{and hence,}$$

$$\big(w(N) - w(S)\big)b^w(S) \ge w(S)b^w(N-S) \quad \text{or equivalently,}$$

$$g^w(S)b^w(N) \ge g^w(N)b^w(S).$$

We have shown $\tau(w) \in C(w)$. Further, $\tau(w) = \tau(v) - d$ by the relative invariance under S-equivalence of the $\tau$-value. From both $\tau(w) = \tau(v) - d$ and $w = v - d$, it is clear that $\tau(w) \in C(w)$ iff $\tau(v) \in C(v)$. It follows that $\tau(v) \in C(v)$. □

The game v of Example 3.7 illustrates that a semiconvex 4-person game which is not zero-monotonic, may possess an empty core. For semiconvex n-person games where $n \geq 5$ we remark that the zero-monotonicity condition is in general not sufficient for the $\tau$-value to belong to the core.

## 7. The $\tau$-value of a quasibalanced simple game

We recall that a player in a simple game is said to be a veto player if and only if any coalition not containing the involved player is losing. In fact, the set $J^v$ of all veto players in a simple game v can also be characterized as the intersection of all winning coalitions in the game.

PROPOSITION 7.1. $J^v = \bigcap\{S \mid S \subset N, v(S) = 1\}$ for all $v \in S^n$.

PROOF. Let $v \in S^n$. Define $\mathcal{W} := \{S \subset N \mid v(S) = 1\}$ and let $i \in N$. It is sufficient to show that $i \notin J$ if and only if there exists $S \in \mathcal{W}$ with $i \notin S$. Clearly, $i \notin J$ implies $v(N-\{i\}) = 1$ and hence, $N-\{i\} \in \mathcal{W}$. It remains to prove the "if part". Suppose that there exists $S \in \mathcal{W}$ with $i \notin S$. Then we have $S \subset N-\{i\}$ such that $v(S) = 1$. Now it follows from the monotonicity condition (1.3) for the simple game v that $v(N-\{i\}) = 1$ and thus, $i \notin J$. □

The upper vector, the gap function and the concession vector of a simple game are described in the next lemma. As a main result, we obtain that the quasibalancedness of a simple game is equivalent to the existence of veto players in the simple game.

LEMMA 7.2. Let $v \in S^n$. Then

(i) $b_i^v = 1$ for all $i \in J^v$

$\quad\; = 0$ for all $i \in N-J^v$.

(ii) $g^v(S) = |J^v| - 1$ whenever $v(S) = 1$

$\quad\; = |S \cap J^v|$ whenever $v(S) = 0$.

In particular, $g^v(S) \geq 0$ for all $S \subset N$ iff $J^v \neq \emptyset$.

(iii) If $J^v \neq \emptyset$, then $g^v(\{i\}) = 0$ for all $i \in N-J^v$.

If $|J^v| \geq 2$, then $g^v(\{i\}) = 1$ for all $i \in J^v$.

(iv) If $|J^v| = 1$, then $g^v(N) = 0$ and $\lambda_i^v = 0$ for all $i \in N$.
If $|J^v| \geq 2$, then $g^v(N) = |J^v| - 1$ and

$$\lambda_i^v = g^v(\{i\}) \text{ for all } i \in N.$$

(v) $v \in QB^n$ iff $J^v \neq \emptyset$.

PROOF. (i) We have $b_i = v(N) - v(N-\{i\}) = 1 - v(N-\{i\})$ for all $i \in N$. Hence, $b_i = 1$ if $i \in J$ and $b_i = 0$ if $i \in N-J$.
(ii) Part (i) implies $g(S) = b(S) - v(S) = |S \cap J| - v(S)$ for all $S \subset N$. Thus, $g(S) = |S \cap J| \geq 0$ whenever $v(S) = 0$. In case $v(S) = 1$, then $J \subset S$ because of Proposition 7.1 and hence, $g(S) = |J| - 1$ whenever $v(S) = 1$. So, (ii) holds.
(iii) Suppose $J \neq \emptyset$. For any $i \in J$ we have $v(N-\{i\}) = 0$ and hence, $v(\{j\}) = 0$ for all $j \in N-\{i\}$ by the monotonicity of the game v. In particular, $v(\{j\}) = 0$ for all $j \in N-J$ and in addition, $v(\{j\}) = 0$ for all $j \in N$ whenever $|J| \geq 2$. Now it follows that $J \neq \emptyset$ implies

$$g(\{j\}) = b_j - v(\{j\}) = b_j = 0 \text{ for all } j \in N-J \quad \text{and further,}$$

$$g(\{j\}) = b_j - v(\{j\}) = b_j = 1 \text{ for all } j \in J \quad \text{when } |J| \geq 2.$$

(iv) Suppose $J \neq \emptyset$. By part (ii), we get $g(N) = |J| - 1$ and $g(S) \geq 0$ for all $S \subset N$. In view of the formula (3.3) and part (iii), we obtain that $\lambda_i = g(\{i\}) = 0$ for all $i \in N-J$.
If $|J| = 1$, then $g(N) = 0$ and hence, $\lambda_i = 0$ for all $i \in N$.
It remains to consider the case $|J| \geq 2$ and $i \in J$. Now it follows from $|J| \geq 2$, part (ii) and part (iii) that

$$g(S) \geq 1 = g(\{i\}) \qquad \text{for all } S \subset N \text{ satisfying } i \in S.$$

From this we conclude that $\lambda_i = g(\{i\}) = 1$ for all $i \in J$ in case $|J| \geq 2$. So, (iv) holds.
(v) From (3.4) and part (ii) we derive that $v \in QB^n$ if and only if $J \neq \emptyset$ and $\lambda(N) \geq g(N)$. However, part (iv) yields

$$\lambda(N) = 0 = g(N) \qquad \text{whenever } |J| = 1,$$

$$\lambda(N) = |J| > |J| - 1 = g(N) \qquad \text{whenever } |J| \geq 2.$$

Therefore, $v \in QB^n$ if and only if $J \neq \emptyset$. So, (v) holds. □

In view of Lemma 7.2(iv), a simple n-person game v with only one veto player possesses a nonnegative gap function $g^v$ with $g^v(N) = 0$ and thus, the game v is 1-convex, while the core $C(v) = \{\tau(v)\} = \{b^v\}$ because of Proposition 3.4.

Further, a simple game with at least two veto players is semiconvex as well as zero-normalized and therefore, the $\tau$-value of the game is proportional to the upper vector by Theorem 6.2(iii). As a consequence, the veto players in a simple game distribute the total savings of the grand coalition equally among themselves according to the $\tau$-value payoff vector for the simple game.

THEOREM 7.3. Let $v \in S^n$ be such that $J^v \neq \emptyset$. Then

$$\begin{aligned} \tau_i(v) &= |J^v|^{-1} && \text{for all } i \in J^v \\ &= 0 && \text{for all } i \in N-J^v. \end{aligned}$$

In particular, $\tau(v) \in C(v)$.

PROOF. It remains to prove $\tau(v) \in C(v)$. Put $x := \tau(v)$ and let $S \subset N$. Clearly, we have $x(S) \geq v(S)$ whenever $v(S) = 0$. In case $v(S) = 1$, then $J \subset S$ because of Proposition 7.1 and hence, $x(S) = 1 = v(S)$. So, $x \in C(v)$ by (2.8). □

COROLLARY 7.4. Let $v \in S^n$. The following three statements are equivalent.

(i) $C(v) \neq \emptyset$. (ii) $v \in QB^n$. (iii) $J^v \neq \emptyset$.

PROOF. We prove (i) ⇒ (ii) ⇒ (iii) ⇒ (i). The implication (i) ⇒ (ii) is a direct consequence of Corollary II.8.5 and Proposition 3.1. The implication (ii) ⇒ (iii) holds because of Lemma 7.2(v), while the implication (iii) ⇒ (i) follows immediately from Theorem 7.3. □

The corollary expresses that the notions of balancedness and quasibalancedness are equivalent on the class of simple n-person games. Moreover, the $\tau$-value of a quasibalanced simple game belongs to the core of the game by Theorem 7.3.

## 8. The $\tau$-value of a game with a nonempty imputation set

In Section 2 we defined the $\tau$-value of a quasibalanced game. In this section we extend the $\tau$-value from the full-dimensional cone $QB^n$ of quasibalanced n-person games to the class $I^n$ of n-person games with a nonempty imputation set. The extension is based on the idea of imposing taxes on the formation of nontrivial coalitions by a multiplicative charge principle. The procedure is described as follows.

Consider a game $v \in G^n$, a nontrivial coalition S and a real number $\epsilon$ satisfying $0 \le \epsilon \le 1$. If there is no cooperation between the members of S, then we suppose that any member i of S acts alone and earns his single worth $v(\{i\})$. In case of cooperation, the coalition S is formed and its worth v(S) can be seen as the total reward to the members of S due to their cooperation. So, the relative reward to the coalition S for its formation from the individuals is given by $v(S) - \sum_{j \in S} v(\{j\})$. Now the idea is to impose a tax on the formation of the coalition S which is proportional to its relative reward. Here the nonnegative real number $\epsilon$ is seen as the tax rate-factor. That is, we impose a cost of

$$\epsilon\left[v(S) - \sum_{j \in S} v(\{j\})\right] \qquad \left(\text{bonus of } -\epsilon\left[v(S) - \sum_{j \in S} v(\{j\})\right]\right.$$

respectively) whenever the relative reward to the coalition S is nonnegative (nonpositive).
According to the above procedure, the induced worth of any nontrivial coalition S is obtained by adding the total earnings of the members of S by noncooperation to the remaining part of the relative reward to S for cooperation.

DEFINITION 8.1. Let $v \in G^n$ and $0 \le \epsilon \le 1$.
The <u>multiplicative</u> <u>$\epsilon$-tax</u> <u>game</u> $v^\epsilon \in G^n$ corresponding to the game v is given by $v^\epsilon(N) := v(N)$ and for all $S \subset N$, $S \ne N$,

$$v^\epsilon(S) := \sum_{j \in S} v(\{j\}) + (1-\epsilon)\left[v(S) - \sum_{j \in S} v(\{j\})\right]. \qquad (3.15)$$

If the game v is zero-normalized and $0 \le \epsilon \le 1$, then the imposed tax on the formation of any nontrivial coalition S is equal to the amount $\epsilon v(S)$ and hence, the induced worth $v^\epsilon(S) = (1-\epsilon)v(S)$. So, for zero-normalized games, the tax is imposed in a multiplicative way with respect to the game itself. The very last observation explains the additional term multiplicative in the above definition.
If $\epsilon = 0$, then no tax is imposed on the formation of any nontrivial coalition and hence, $v^0 = v$. If $\epsilon = 1$, then the imposed tax on the formation of any nontrivial coalition equals its total relative reward and hence, the game $v^1$ is given by $v^1(N) = v(N)$ and

$$v^1(S) = \sum_{j \in S} v(\{j\}) \qquad \text{for all } S \subset N,\ S \ne N. \qquad (3.16)$$

In case $v \in I^n$ where $n \geq 2$, then the corresponding game $v^1$ is also called the induced <u>bargaining game</u> since the game $v^1$ is characterized by the fact that the grand coalition N is the only multiperson coalition for which agreement on cooperation is interesting. Obviously, $I(v^1) = I(v) = C(v^1)$ and from this, it follows that the induced bargaining game $v^1$ is always balanced as well as quasibalanced, although the original game $v \in I^n$ may not be (quasi)balanced. In order to extend the $\tau$-value from $QB^n$ to $I^n$, we determine for any game $v \in I^n$ its corresponding least quasibalanced multiplicative tax game where the rate-factor is minimal in the sense that a smaller tax rate-factor gives rise to a multiplicative tax game that is not quasibalanced any more. Formally, for any game $v \in I^n$ we define the critical number

$$\epsilon^{qb}(v) := \min\left(\epsilon \mid \epsilon \geq 0,\ v^{\epsilon} \in QB^n\right). \tag{3.17}$$

LEMMA 8.2. Let $v \in I^n$.
(i) The critical number $\epsilon^{qb}(v)$ is well-defined.
(ii) $0 \leq \epsilon^{qb}(v) \leq 1$ and $\epsilon^{qb}(v) = 0$ iff $v \in QB^n$.

PROOF. By the two formulas (3.15) and (3.16), we have that $v^{\epsilon} = v^1 + (1-\epsilon)(v - v^1)$ for all $0 \leq \epsilon \leq 1$. Further, $QB^n$ is a polyhedral cone by Proposition 3.2. Thus, the class $QB^n$ is a closed convex set of n-person games. Notice that $v^1 \in QB^n$. Now it follows that the critical number $\epsilon^{qb}(v)$ of (3.17) is well-defined. The statement (ii) becomes clear by noting that $v^1 \in QB^n$ and $v^0 = v$. □

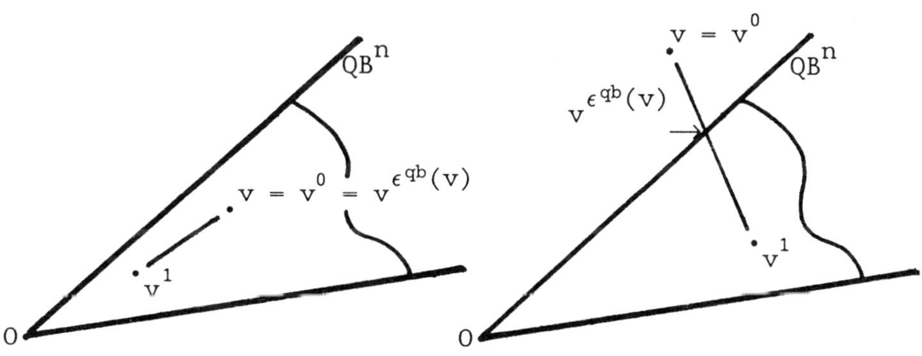


FIGURE 3.3. The geometric position of the tax game $v^{\epsilon^{qb}(v)}$ within the polyhedral cone $QB^n$ in case $v \in QB^n$ and $v \notin QB^n$ respectively.

Figure 3.3 illustrates the geometric position of the least quasibalanced multiplicative tax game $v^{\epsilon^{qb}(v)}$ with respect to the original n-person game v and the cone $QB^n$ as well.

The corresponding least quasibalanced multiplicative tax game will be used to define the $\tau$-value of a game with no dummy players. For any game with at least one dummy we first remove the dummies in order to meet the dummy player property for values. Here the set of all dummy players in a game $v \in G^n$ is denoted by $D^v$ or shortly D. Thus,

$$D^v := \{i \in N \mid v(S \cup \{i\}) - v(S) = v(\{i\}) \text{ for all } S \subset N-\{i\}\}.$$

Notice that the subgame $(N-D^v; v_{N-D})$ is obtained by removing the dummies in the given game v and there are no dummy players in this induced subgame. The relationships between the upper vectors, the gap functions and the concession vectors respectively of the original game and the induced subgame are described with the aid of the next two lemmas.

LEMMA 8.3. Let $v \in G^n$. Then

(i) $v(S) = v(S-D^v) + \sum_{j \in S \cap D^v} v(\{j\})$ for all $S \subset N$. (3.18)

(ii) If $D^v = N$, then $v(S) = \sum_{j \in S} v(\{j\})$ for all $S \subset N$,

$v \in QB^n$ and $\tau_i(v) = v(\{i\})$ for all $i \in N$.

PROOF. (i) Let $S \subset N$. We first note that (3.18) trivially holds whenever $S \cap D = \emptyset$. Thus, we may suppose $S \cap D \neq \emptyset$. Write $S \cap D = \{i_1, i_2, \ldots, i_k\}$ where $k \geq 1$. Put $T := S-D$, $[i_0] := \emptyset$ and $[i_j] := \{i_1, i_2, \ldots, i_j\}$ for all $1 \leq j \leq k$. Now it follows that

$$v(S) - v(S-D) = \sum_{j=1}^{k} \big(v(T \cup [i_j]) - v(T \cup [i_{j-1}])\big)$$

$$= \sum_{j=1}^{k} \big(v(T \cup [i_{j-1}] \cup \{i_j\}) - v(T \cup [i_{j-1}])\big)$$

$$= \sum_{j=1}^{k} v(\{i_j\}) = \sum_{j \in S \cap D} v(\{j\}).$$ So, (i) holds.

The statement (ii) is a direct consequence of part (i) and the dummy player property of the $\tau$-value on the cone $QB^n$. □

LEMMA 8.4. Let $v \in G^n$ satisfy $D^v \neq N$ and let $w = v_{N-D}$. Then

(i) $b_i^v = v(\{i\})$ for all $i \in D^v$,

$b_i^v = b_i^w$ for all $i \in N-D^v$.

(ii) $g^v(S) = g^w(S-D^v)$ for all $S \subset N$.

(iii) $\lambda_i^v \leq 0$ for all $i \in D^v$,

$\lambda_i^v = \lambda_i^w$ for all $i \in N-D^v$.

PROOF. Put $M := N-D^v$.

(i) Clearly, $b_i^v = v(N) - v(N-\{i\}) = v(\{i\})$ whenever $i \in D^v$. For all $j \in M$, it follows from $(N-\{j\}) \cap D^v = D^v$ and (3.18) that

$$b_j^v = v(N) - v(N-\{j\}) = v(M) - v(M-\{j\})$$
$$= w(M) - w(M-\{j\}) = b_j^w.$$

(ii) From part (i) and (3.18) we derive that for all $S \subset N$

$$g^v(S) = b^v(S) - v(S) = b^w(S-D^v) + \sum_{j \in S \cap D^v} v(\{j\}) - v(S)$$
$$= b^w(S-D^v) - v(S-D^v) = b^w(S-D^v) - w(S-D^v) = g^w(S-D^v).$$

(iii) For all $i \in D^v$ we have $\lambda_i^v \leq g^v(\{i\}) = b_i^v - v(\{i\}) = 0$ by using (3.3), (3.2) and part (i) respectively. Let $j \in M$. From $\{S \subset M |\ j \in S\} \subset \{S \subset N |\ j \in S\}$ as well as part (ii), it follows that

$$\lambda_j^v = \min[g^v(S) |\ S \subset N,\ j \in S]$$
$$\leq \min[g^v(S) |\ S \subset M,\ j \in S]$$
$$= \min[g^w(S) |\ S \subset M,\ j \in S] = \lambda_j^w.$$

Further, choose $T \subset N$ such that $j \in T$ and $g^v(T) = \lambda_j^v$. Then $j \in T-D^v$ and part (ii) imply $\lambda_j^w \leq g^w(T-D^v) = g^v(T) = \lambda_j^v$. We conclude that $\lambda_j^v = \lambda_j^w$ for all $j \in M$. □

The next theorem states that the payoffs to the nondummy players in a quasibalanced game are not affected whenever the dummy players are removed, on the understanding that the players are paid according to the $\tau$-value concept.

THEOREM 8.5. Let $v \in QB^n$ be such that $D^v \neq N$. Then the induced subgame $(N-D^v; v_{N-D})$ is also quasibalanced and

$$\tau_i(v_{N-D}) = \tau_i(v) \text{ for all } i \in N-D^v.$$

PROOF. Put $w := v_{N-D}$ and $M := N-D^v$. Notice that $\lambda_i^v \geq 0$ for all $i \in N$ because $v \in QB^n$. By means of Lemma 8.4, we obtain

$$g^v(N) = g^w(M), \qquad g^v(S) = g^w(S) \qquad \text{for all } S \subset M,$$
$$b_i^v = b_i^w, \qquad \lambda_i^v = \lambda_i^w \qquad \text{for all } i \in M,$$
$$\text{as well as} \qquad \lambda_i^v = 0 \qquad \text{for all } i \in D^v.$$

Particularly, $\lambda^w(M) = \lambda^v(N)$. In view of (3.4) and $v \in QB^n$, it is now obvious that the game $w$ is also quasibalanced. For all $j \in M$ we derive from the known formula (3.6) that

$$\tau_j(w) = b_j^w = b_j^v = \tau_j(v) \qquad \text{whenever } g^v(N) = 0 \text{ and}$$
$$\tau_j(w) = b_j^w - g^w(M)\left(\lambda^w(M)\right)^{-1}\lambda_j^w$$
$$= b_j^v - g^v(N)\left(\lambda^v(N)\right)^{-1}\lambda_j^v = \tau_j(v) \quad \text{whenever } g^v(N) > 0. \qquad \square$$

Now we present a well-defined extension of the $\tau$-value from the class $QB^n$ of quasibalanced n-person games to the class $I^n$ of n-person games with a nonempty imputation set.

DEFINITION 8.6.
The <u>$\tau$-value</u> $\tau(v) \in \mathbb{R}^n$ of a game $v \in I^n$ satisfying $D^v = \emptyset$ is <u>defined to</u> be the $\tau$-value of the corresponding least quasibalanced multiplicative tax game $v^{\epsilon^{qb}(v)} \in QB^n$, i.e.,

$$\tau(v) := \tau(v^{\epsilon^{qb}(v)}).$$

The <u>$\tau$-value</u> $\tau(v) \in \mathbb{R}^n$ of a game $v \in I^n$ satisfying $D^v \neq N, \emptyset$, is given by

$$\tau_i(v) := v(\{i\}) \qquad \text{for all } i \in D^v$$
$$= \tau_i(v_{N-D}) \qquad \text{for all } i \in N-D^v.$$

The $\tau$-value of any game $v \in I^n$ with no dummy players is determined by the $\tau$-value of the specifically chosen quasibalanced n-person game $v^{\epsilon^{qb}(v)}$. For games $v \in I^n$ with a nontrivial dummy player set, the $\tau$-value payoff to any dummy player equals the worth of his own coalition, while the $\tau$-value payoff to the nondummy players in the game v is determined according to the $\tau$-value payoff in the induced dummy free subgame $(N-D^v;v_{N-D})$.

In view of the dummy player property of the $\tau$-value on $QB^n$, Lemma 8.2(ii) and Theorem 8.5, we conclude that the $\tau$-value on $I^n$ so defined can indeed be seen as an extension of the $\tau$-value on $QB^n$. As a matter of fact, the extended $\tau$-value possesses the standard properties (i)-(iv) for values as mentioned in Section II.1.

THEOREM 8.7. The $\tau$-value $\tau$: $I^n \to \mathbb{R}^n$ possesses the following five properties: (i) individual rationality; (ii) symmetry; (iii) dummy player property; (iv) relative invariance under S-equivalence and (v) substitution property.

PROOF. Let $v \in I^n$. By Lemma 8.3(ii), we have that $v \in QB^n$ when $D^v = N$. Thus, we may suppose $D^v \neq N$. Put $w := v^{\epsilon^{qb}(v)}$. Then we get $w \in QB^n$ and therefore, the standard properties of the $\tau$-value on $QB^n$ apply to the game w.
(0) We first treat the efficiency property of $\tau$. If $D^v = \emptyset$, then
$\tau(v) = \tau(w)$ and hence, $\sum_{j\in N} \tau_j(v) = \sum_{j\in N} \tau_j(w) = w(N) = v(N)$.

If $D^v \neq N,\emptyset$, then $\sum_{j\in N} \tau_j(v) = \sum_{j\in N-D^v} \tau_j(v_{N-D}) + \sum_{j\in D^v} v(\{j\})$

$= v_{N-D}(N-D^v) + \sum_{j\in D^v} v(\{j\}) = v(N-D^v) + \sum_{j\in D^v} v(\{j\}) = v(N)$

where the second equality follows from the efficiency property of $\tau$ applied to dummy free games and the fourth equality from the formula (3.18). So, the $\tau$-value on $I^n$ is indeed efficient.
(i) We have $\tau_i(v) = \tau_i(w) \geq w(\{i\}) = v(\{i\})$ for all $i \in N$ whenever $D^v = \emptyset$. In case $D^v \neq N,\emptyset$, then we obtain that

$$\tau_i(v) = v(\{i\}) \qquad \text{for all } i \in D^v,$$

$$\tau_i(v) = \tau_i(v_{N-D}) \geq v_{N-D}(\{i\}) = v(\{i\}) \text{ for all } i \in N-D^v$$

where the inequality follows from the individual rationality of $\tau$ applied to dummy free games. So, the $\tau$-value on $I^n$ is individually rational.

(ii) Let $\theta: N \to N$ be a permutation. Then the equivalence $i \in D^v$ iff $\theta(i) \in D^{\theta v}$ holds. By straightforward calculations using the two formulas (3.15) and (3.17), we obtain that

$$(\theta v)^{\epsilon} = \theta(v^{\epsilon}) \qquad \text{for all } 0 \le \epsilon \le 1 \qquad \text{and hence,}$$

$$\epsilon^{qb}(\theta v) = \epsilon^{qb}(v).$$

If $D^v = \emptyset$, then $D^{\theta v} = \emptyset$ and it follows that for all $i \in N$

$$\tau_{\theta(i)}(\theta v) = \tau_{\theta(i)}\left[(\theta v)^{\epsilon^{qb}(\theta v)}\right] = \tau_{\theta(i)}\left[(\theta v)^{\epsilon^{qb}(v)}\right]$$

$$= \tau_{\theta(i)}\left[\theta(v^{\epsilon^{qb}(v)})\right] = \tau_{\theta(i)}(\theta w) = \tau_i(w) = \tau_i(v).$$

In case $D^v \neq N,\emptyset$, then $D^{\theta v} \neq N,\emptyset$, and it follows that

$$\tau_{\theta(i)}(\theta v) = (\theta v)(\{\theta(i)\}) = v(\{i\}) = \tau_i(v) \quad \text{for } i \in D^v,$$

$$\tau_{\theta(i)}(\theta v) = \tau_{\theta(i)}\left[(\theta v)_{N-D^{\theta v}}\right] = \tau_{\theta(i)}\left[\theta(v_{N-D^v})\right]$$

$$= \tau_i\left[v_{N-D^v}\right] = \tau_i(v) \qquad \text{for all } i \in N-D^v$$

where the last equality but one follows from the symmetry property of $\tau$ applied to dummy free games. So, the $\tau$-value on $I^n$ is symmetric.

(iii) The dummy player property of the $\tau$-value on $I^n$ is obvious because $\tau_i(v) = v(\{i\})$ for all $i \in D^v$.

(iv) Let $\alpha \in (0,\infty)$ and $d \in \mathbb{R}^n$. Then $i \in D^v$ iff $i \in D^{\alpha v+d}$. Further, in a straightforward way we derive from the two formulas (3.15) and (3.17) that

$$(\alpha v+d)^{\epsilon} = \alpha v^{\epsilon} + d \qquad \text{for all } 0 \le \epsilon \le 1 \qquad \text{and hence,}$$

$$\epsilon^{qb}(\alpha v+d) = \epsilon^{qb}(v).$$

If $D^v = \emptyset$, then $D^{\alpha v+d} = \emptyset$ and it follows that

$$\tau(\alpha v + d) = \tau\left[(\alpha v+d)^{\epsilon^{qb}(\alpha v+d)}\right] = \tau\left[(\alpha v+d)^{\epsilon^{qb}(v)}\right]$$

$$= \tau\left[\alpha v^{\epsilon^{qb}(v)} + d\right] = \tau(\alpha w + d) = \alpha\tau(w) + d = \alpha\tau(v) + d.$$

In case $D^v \neq N,\emptyset$, then $D^{\alpha v+d} \neq N,\emptyset$, and it follows that

$$\tau_i(\alpha v + d) = (\alpha v+d)(\{i\}) = \alpha v(\{i\}) + d_i = \alpha\tau_i(v) + d_i$$

for all $i \in D^v$,

while we have that for all $i \in N-D^v$

$$\tau_i(\alpha v + d) = \tau_i\left((\alpha v+d)_{N-D^{\alpha v+d}}\right) = \tau_i\left((\alpha v+d)_{N-D^v}\right)$$

$$= \tau_i\left(\alpha v_{N-D^v} + d\right) = \alpha\tau_i\left(v_{N-D^v}\right) + d_i = \alpha\tau_i(v) + d_i$$

where the last equality but one follows from the relative invariance under S-equivalence of $\tau$ applied to dummy free games. So, the $\tau$-value on $I^n$ is relatively invariant under S-equivalence.

(v) Let $i \in N$ and $j \in N$ be substitutes in the game v. That is, $v(S \cup \{i\}) = v(S \cup \{j\})$ for all $S \subset N-\{i,j\}$. Thus, we have $v(\{i\}) = v(\{j\})$. Further, in a straightforward way we obtain that the two players i and j are also substitutes in the multiplicative tax game w. In addition, the equivalence $i \in D^v$ iff $j \in D^v$ holds.

We deduce $\tau_i(v) = \tau_i(w) = \tau_j(w) = \tau_j(v)$ whenever $D^v = \emptyset$.

In case $D^v \neq N,\emptyset$, then it follows that

$$\tau_i(v) = v(\{i\}) = v(\{j\}) = \tau_j(v) \qquad \text{whenever } i \in D^v,$$

$$\tau_i(v) = \tau_i(v_{N-D}) = \tau_j(v_{N-D}) = \tau_j(v) \qquad \text{whenever } i \in N-D^v$$

where the last equality but one follows from the substitution property of $\tau$ applied to dummy free games. So, the $\tau$-value on $I^n$ possesses the substitution property. □

For the moment, Theorem 8.7 concludes the main part of the solution theory based on the $\tau$-value concept. For minor results concerning the $\tau$-value notion and related subjects, we refer to Tijs (1981, 1987); Tijs and Lipperts (1982); Driessen and Tijs (1983, 1984, 1985a); Tijs and Driessen (1986b, 1987); Driessen (1987b). Finally, the next example illustrates the computation of the $\tau$-value for a game which is not quasibalanced.

EXAMPLE 8.8. Let the zero-normalized 3-person game v be given by

$$v(12) = 75, \quad v(13) = 85, \quad v(23) = 90 \quad \text{and} \quad v(N) = 100.$$

Then $D^v = \emptyset$ and $b^v = (10,15,25)$. Note that $v \notin QB^3$ because $g^v(N) = -50 < 0$. For any $0 \leq \epsilon \leq 1$, the corresponding zero-normalized multiplicative $\epsilon$-tax 3-person game $v^\epsilon$ is given by

$$v^{\epsilon}(12) = 75(1-\epsilon), \qquad v^{\epsilon}(13) = 85(1-\epsilon),$$
$$v^{\epsilon}(23) = 90(1-\epsilon), \qquad v^{\epsilon}(N) = 100.$$

In particular, the induced bargaining game $v^1$ is given by

$$v^1(N) = 100 \quad \text{and} \quad v^1(S) = 0 \text{ for all } S \subset N,\ S \neq N.$$

We have $b^{v^{\epsilon}} = (10+90\epsilon, 15+85\epsilon, 25+75\epsilon)$ and $g^{v^{\epsilon}}(N) = 250\epsilon-50$. We get $g^{v^{\epsilon}}(N) \geq 0$ iff $0.2 \leq \epsilon \leq 1$. Now it follows that

$$\epsilon^{qb}(v) = 0.2 \quad \text{and} \quad \tau(v) = \tau(v^{0.2}) = (28,32,40).$$

It is left to the reader to verify that the Shapley value and the nucleolus payoff vector for the game v are given by

$$\phi(v) = \frac{1}{2}(60,65,75) \quad \text{and} \quad \eta(v) = \frac{1}{3}(80,95,125).$$

In view of the vector equality $\tau(v) = 0.4\phi(v) + 0.6\eta(v)$, we observe that the $\tau$-value of the game v lies on the straight line segment with end points the Shapley value and the nucleolus of the game v. Here the $\tau$-value payoffs to the two players 1 and 2 are more than their nucleolus payoffs, but less than their Shapley value payoffs. In comparison with his $\tau$-value payoff, the third player gets more by the nucleolus concept, but less by the Shapley value concept. As we shall see in the next chapter, these three one-point solution concepts are useful in a game theoretic analysis of the cost allocation problem.

CHAPTER IV

# A GAME THEORETIC APPROACH TO THE COST ALLOCATION PROBLEM BY MEANS OF THE $\tau$-VALUE, THE NUCLEOLUS AND THE SHAPLEY VALUE

## 1. The TVA cost allocation problem (Ransmeier, 1942)

In the cost allocation problem which confronted the Tennessee Valley Authority (TVA), two major types of joint project investment cost were distinguished: "direct" costs which are traceable to particular purposes (these costs may be in the nature of marginal costs) and the remaining joint costs.

TVA research into the problem of joint cost allocation took place in several stages. In the first stage (1933-1934), the equal apportionment basis for allocation of Wilson Dam joint investment cost was favoured. With respect to Wilson Dam, no charge to flood control and fertilizer production was considered justified. As an initial step, direct costs were charged to the remaining three project objectives (navigation, power and national defense as well), while it was considered fair to divide the remaining joint costs (determined by the present value of Wilson Dam minus the sum of direct costs) equally between navigation and power since the associated single purpose costs (which consists of constructing a dam solely for the development of that objective) are equal. However, the TVA Board refrained from adopting the equal charge basis of allocation at Wilson Dam. The main reason for this may have been that generally speaking the equal charge principle fails to consider the justifiability of the charges it assesses.

In the second stage of TVA allocation research (1935-1937), the "benefit" allocation method was developed. For each project objective, its remaining benefit is determined by subtracting its direct cost from its benefit and then joint costs are divided in the proportions of remaining benefits. Early benefit allocation methods employed estimated costs of alternate single purpose systems as the fundamental criteria of benefits. The so-called alternate cost avoided allocation method (1938) which will be treated in the next section, is an example of a benefit allocation method on the basis of simple alternate costs. These early benefit allocation methods were deficient in that they failed to investigate the justifiability of the estimated alternate expenditures.

In the third and final stage of TVA allocation research (1937-1942), the "alternative justifiable expenditure" allocation method was developed. This method is a version of benefit theory according to which the involved costs of alternate single purpose systems or alternative justifiable expenditures, whichever are less, are taken as measures of benefits. Although the TVA has never officially adopted any single basis of cost allocation, it has been primarily influenced by the above method. For a closer study of the TVA cost allocation problem we refer to Ransmeier (1942).

## 2. The $\tau$-value in comparison with cost allocation methods based on separable and nonseparable costs

The purpose of this section is to show that, for some subclasses of cost games, the cost allocation proposed by the $\tau$-value concept agrees with the cost apportionment by well-known cost allocation methods.

Let $N = \{1,2,\ldots,n\}$ be a set of users who cooperate in the undertaking of a joint project and, for any nonempty subset $S$ of $N$, let $c(S)$ represent the least cost of undertaking a similar joint venture which only serves the members of the involved coalition $S$. By putting $c(\emptyset) := 0$, the resulting cost function $c\colon 2^N \to \mathbb{R}$ must be <u>subadditive</u>, i.e.,

$$c(S) + c(T) \geq c(S \cup T)$$
for any two disjoint coalitions $S$ and $T$

since the ways of serving $S$ together with $T$ include the possibility of serving $S$ alone and $T$ alone. The above cost function $c$ is regarded as the characteristic cost function of the <u>cost game</u> $(N;c)$. With a given cost game $(N;c)$ we associate a <u>savings game</u> $(N;v)$ by means of

$$v(S) := \sum_{j\in S} c(\{j\}) - c(S) \qquad \text{for all } S \subset N. \tag{4.1}$$

The worth $v(S)$ of a coalition $S$ represents the cost savings that would result in the cost game from cooperation between the members of $S$ instead of acting alone. The subadditivity of the cost function $c$ implies that the associated savings function $v$ is nonnegative and superadditive. Furthermore,

the savings game is always zero-normalized because there are no cost savings for single users.
A vector $y \in \mathbb{R}^n$ is called a <u>cost allocation</u> for a cost game (N;c) if it satisfies the efficiency principle in terms of joint costs, i.e., $y(N) = c(N)$. Here the i-th coordinate $y_i$ of the cost allocation y represents the cost allocated to user i. The cost allocation methods prescribe somehow a specific (one-point) subset of the cost allocation set. With any cost allocation y for a cost game c we associate an efficient payoff vector x for the corresponding savings game v by means of

$$x_i := c(\{i\}) - y_i \qquad \text{for all } i \in N. \text{ Then } x \in I(v) \text{ iff}$$
$$y_i \leq c(\{i\}) \qquad \text{for all } i \in N \text{ and } y(N) = c(N).$$

Traditional inequalities in game theory have to be reversed whenever we are dealing with costs instead of savings.

A principle with firm roots in the multipurpose water resource project evaluation literature is that no user should be charged less than the separable cost of including the user in the joint project (Federal Inter-Agency River Basin Committee, 1950; Inter-Agency Committee on Water Resources, 1958; Water Resources Council, 1962).

DEFINITION 2.1. Let (N;c) be a cost game.
The <u>separable</u> (or <u>marginal</u>) <u>cost</u> $SC_i(c)$ <u>of player</u> $i \in N$ and the <u>nonseparable</u> <u>cost</u> NSC(c) in the cost game c are given by

$$SC_i(c) := c(N) - c(N-\{i\}) \qquad \text{for all } i \in N, \tag{4.2}$$

$$NSC(c) := c(N) - \sum_{j \in N} SC_j(c). \tag{4.3}$$

Separable costs are equal to the total change in the cost of a project due to adding an user to the project already designed for all other users. Separable costs include direct costs (which are solely traceable to single users) and the additional costs of changing the size of the multi-group cost elements. As such, the use of separable costs represents a significant change from the TVA procedures which used direct costs.
Note that the formula (4.2) and the subadditivity of the cost function c imply that the separable cost of any user is at most his cost of acting alone in the cost game, i.e.,

$SC_i(c) \le c(\{i\})$ for all $i \in N$. Given that the users are already charged their separable costs, the remaining cost to be assigned is the nonseparable cost. Hence, the cost allocation problem becomes the problem of how to allocate the nonseparable cost. Generally, each user's prorated share of the nonseparable cost can be described by means of some vector $\alpha(c) \in \mathbb{R}^n$ whose coordinates sum up to one, where the vector may depend upon the involved cost game c. The coordinates of the corresponding vector are regarded as (usually nonnegative) weights which determine the fractions of the nonseparable cost allocated to the users in the cost game. Next we consider three allocation methods based on separable and nonseparable costs.

DEFINITION 2.2. Let $CG^n$ be the class of n-person cost games and let G be any nonempty collection of cost games in $CG^n$. A <u>separable cost allocation method on</u> G is a map $M: G \to \mathbb{R}^n$ such that for all $c \in G$ and all $i \in N$

$$M_i(c) = SC_i(c) + \alpha_i(c)NSC(c)$$

where $\alpha(c) \in \mathbb{R}^n$ is any n-tuple satisfying $\sum_{j=1}^{n} \alpha_j(c) = 1$.

A separable cost allocation method $M: G \to \mathbb{R}^n$ is called

(1) the <u>egalitarian nonseparable cost</u> (ENSC-) <u>method</u> if

$$\alpha_i(c) = n^{-1} \quad \text{for all } c \in G \text{ and all } i \in N. \tag{4.4}$$

(2) the <u>alternate cost avoided</u> (ACA-) <u>method</u> whenever for all $c \in G$ and all $i \in N$

$$\alpha_i(c) = \beta_i(c)\Big(\sum_{j \in N} \beta_j(c)\Big)^{-1} \quad \text{where}$$

$$\beta_i(c) := c(\{i\}) - SC_i(c). \tag{4.5}$$

(3) the <u>separable costs remaining benefits</u> (SCRB-) <u>method</u> if for all $c \in G$ and all $i \in N$

$$\alpha_i(c) = \gamma_i(c)\Big(\sum_{j \in N} \gamma_j(c)\Big)^{-1} \quad \text{where}$$

$$\gamma_i(c) := \min\big(b_i^*(c), c(\{i\})\big) - SC_i(c).$$

Here $b_i^*(c)$ denotes the benefit to player i in the cost game c by acting alone.

By the ENSC-method, the nonseparable cost is allocated equally to the users. By the ACA-method, the nonseparable cost in a given cost game c is prorated in proportion to the part of the alternate cost $c(\{i\})$ of any user i that can be avoided by including user i in the joint project, i.e., $c(\{i\}) - SC_i(c)$. A first version of the ACA-method (using direct costs instead of separable costs) has been proposed by the TVA consultant Martin Glaeser in 1938. By the SCRB-method, the nonseparable cost is allocated in proportion to each user's willingness to pay minus the separable cost already allocated. Here each user i is not willing to pay more than his benefit $b_i^*(c)$ or his alternate cost $c(\{i\})$ to participate in the joint project. If the benefit of any user exceeds his alternate cost, then the SCRB-method agrees with the ACA-method. Therefore, the SCRB-method is a modification of the ACA-method. The SCRB-method was recommended to the United States Inter-Agency Committee on Water Resources in 1950 (Federal Inter-Agency River Basin Committee, 1950) and is today the most widely used approach in the water resources field.
The next theorem states that the ENSC-method (ACA-method respectively) agrees with the well-known $\tau$-value concept when we are dealing with cost games which generate 1-convex (semiconvex) savings games. In order to prove this theorem, we first relate the separable and nonseparable costs in a cost game to the upper vector and the gap function of the associated savings game. In fact, the nonseparable cost is equal to the gap of the grand coalition, while the gap of a single user equals the part of his alternate cost avoided by including the user in the joint project. The resulting relations are useful to reformulate the 1-convexity and the semiconvexity respectively of the characteristic savings function v in terms of the characteristic cost function c.

LEMMA 2.3. Let $v \in G^n$ be the savings game corresponding to the cost game $c \in CG^n$. Then

$$g^v(\{i\}) = b_i^v = c(\{i\}) - SC_i(c) \qquad \text{for all } i \in N,$$

$$g^v(N) = NSC(c) \qquad \text{and further,}$$

$$g^v(S) = c(S) - \sum_{j \in S} SC_j(c) \qquad \text{for all } S \subset N.$$

PROOF. It follows from (3.1)-(3.2) and (4.1)-(4.2) that

$$b_i^v = v(N) - v(N-\{i\}) = c(\{i\}) - c(N) + c(N-\{i\})$$

$$= c(\{i\}) - SC_i(c) \qquad \text{for all } i \in N \quad \text{and therefore,}$$

$$g^v(S) = b^v(S) - v(S) = \sum_{j\in S} c(\{j\}) - \sum_{j\in S} SC_j(c) - v(S)$$

$$= c(S) - \sum_{j\in S} SC_j(c) \qquad \text{for all } S \subset N. \quad \text{In particular,}$$

$$g^v(N) = c(N) - \sum_{j\in N} SC_j(c) = NSC(c) \text{ because of (4.3).} \ \square$$

THEOREM 2.4. Let $v \in G^n$ be the savings game corresponding to the cost game $c \in CG^n$.

(i) If $c(N) \le c(S) + \sum_{j\in N-S} SC_j(c)$ for all $S \subset N$, $S \ne \emptyset$,

and $NSC(c) \ge 0$, (4.6)

then $\tau_i(v) = c(\{i\}) - ENSC_i(c)$ for all $i \in N$.

(ii) If $c(\{i\}) + \sum_{j\in S-\{i\}} SC_j(c) \le c(S)$ (4.7)

for all $i \in N$ and all $S \subset N$ such that $i \in S$,

then $\tau_i(v) = c(\{i\}) - ACA_i(c)$ for all $i \in N$.

PROOF. By Lemma 2.3, we have both $g^v(N) = NSC(c)$ and

$$b_i^v = c(\{i\}) - SC_i(c) \qquad \text{for all } i \in N. \tag{4.8}$$

(i) In view of (3.11) and Lemma 2.3, it is obvious that the condition (4.6) for the cost game c is equivalent to the 1-convexity condition (3.11) for the corresponding savings game v. It follows from Theorem III.5.3, (4.8) and (4.4) respectively that

$$\tau_i(v) = b_i^v - n^{-1}g^v(N) = c(\{i\}) - SC_i(c) - n^{-1}NSC(c)$$

$$= c(\{i\}) - ENSC_i(c) \qquad \text{for all } i \in N.$$

(ii) By Lemma 2.3, the condition (4.7) is equivalent to $g^v(\{i\}) \le g^v(S)$ for all $i \in N$ and all $S \subset N$ with $i \in S$. Moreover, $g^v(\{i\}) = c(\{i\}) - SC_i(c) \ge 0$ for all $i \in N$. In view of (3.13), the savings game v is semiconvex. In case $g^v(N) = 0$, then also $NSC(c) = 0$ because of (4.8) and thus,

$$\tau_i(v) = b_i^v = c(\{i\}) - SC_i(c) = c(\{i\}) - ACA_i(c)$$

for all $i \in N$. It remains to consider the case $g^v(N) > 0$. Recall that any savings game is superadditive as well as zero-normalized. Now it follows from Theorem III.6.2(iii), (4.5) and (4.8) that

$$\tau_i(v) = v(N)\left(b^v(N)\right)^{-1}b_i^v = b_i^v - g^v(N)\left(b^v(N)\right)^{-1}b_i^v$$

$$= c(\{i\}) - ACA_i(c) \qquad \text{for all } i \in N. \qquad \square$$

For any cost allocation y in a cost game c we consider its corresponding pre-imputation x in the associated savings game v. Then we have that $x_i = c(\{i\}) - y_i$ for all $i \in N$ and in particular, we obtain the result that for all $i \in N$

$$v(\{i\}) \le x_i \le b_i^v \qquad \text{iff} \qquad SC_i(c) \le y_i \le c(\{i\}).$$

Hence, the separable costs (alternate costs respectively) can be seen as minimal (maximal) charges to be paid by the users. By the subadditivity of the cost function c, we get $c(N) \le c(S) + \sum_{j \in N-S} c(\{j\})$ for all nonempty coalitions S. It may happen that these inequalities even hold whenever the maximal charges $c(\{j\})$, $j \in N-S$, are replaced by their minimal charges $SC_j(c)$, $j \in N-S$. In other words, if the joint costs c(N) are allocated in such a way that all users outside a nonempty coalition S pay their minimal charges, then the remaining joint costs are at most the cost of the involved coalition S. Under these circumstances together with a nonnegative nonseparable cost, the $\tau$-value agrees with the ENSC-method in a natural way by Theorem 2.4(i).
It may also happen that the remaining cost for any member of a coalition is at least his maximal charge whenever the other members of the coalition contribute their minimal charges to the cost of the formed coalition. Under these circumstances, the $\tau$-value concept equals the ACA-method in a natural way according to Theorem 2.4(ii).

Finally, we treat the cost allocation problem for the TVA ten dam system in order to illustrate the comparison between the $\tau$-value concept and separable cost allocation methods. The three objectives of navigation, flood control

| coalition S | cost c(S) | savings v(S) |
|---|---|---|
| {1} | 163,520 | 0 |
| {2} | 140,826 | 0 |
| {3} | 250,096 | 0 |
| {1,2} | 301,607 | 2,739 |
| {1,3} | 378,821 | 34,795 |
| {2,3} | 367,370 | 23,552 |
| {1,2,3} | 412,584 | 141,858 |

TABLE 4.1. Cost and saving figures in $ 1000.

| objective i | 1 | 2 | 3 | total |
|---|---|---|---|---|
| separable cost $SC_i(c)$ | 45,214 | 33,763 | 110,977 | 189,954 |
| $c(\{i\}) - SC_i(c)$ $(= g^v(\{i\}))$ | 118,306 | 107,063 | 139,119 | 364,488 |
| allocation to objective i of NSC(c) by the ACA-method | 72,262 | 65,394 | 84,974 | 222,630 |

TABLE 4.2. Calculations for the determination of the cost allocation by the ACA-method (costs in $ 1000). Notice that $g^v(N) = NSC(c) = 222{,}630$.

| objective i | $ENSC_i(c)$ | $ACA_i(c)$ | $c(i)-\tau_i(v)$ | $c(i)-\eta_i(v)$ | $c(i)-\phi_i(v)$ |
|---|---|---|---|---|---|
| 1 | 119,424 (28.9%) | 117,476 (28.5%) | 117,476 (28.5%) | 116,234 (28.2%) | 117,829 (28.6%) |
| 2 | 107,973 (26.2%) | 99,157 (24.0%) | 99,157 (24.0%) | 93,540 (22.7%) | 100,756 (24.4%) |
| 3 | 185,187 (44.9%) | 195,951 (47.5%) | 195,951 (47.5%) | 202,810 (49.1%) | 193,999 (47.0%) |

TABLE 4.3. Cost allocations by the ENSC-, the ACA-method, the $\tau$-value $\tau$, the nucleolus $\eta$ and the Shapley value $\phi$ (costs in $ 1000).

TABLES 4.1-4.3 are related to the cost allocation problem for the TVA ten dam system. The objectives of navigation, flood control and power are denoted as 1, 2, and 3 respectively.

and power are denoted as 1, 2, and 3 respectively. Here the relevant cost figures and the corresponding saving figures are listed in Table 4.1 which is adapted from Ransmeier (1942, page 329). Table 4.2 shows the calculations which are necessary for the determination of the allocation by the alternate cost avoided method. Table 4.3 compares the allocations by two known TVA methods, the ACA- as well as the ENSC-method, with the cost allocations corresponding to the $\tau$-value, the nucleolus and the Shapley value. The cost allocations by both the ACA-method and the $\tau$-value coincide because the cost game generates a semiconvex savings game as can be observed in Table 4.2. The reader is invited to study the three Tables 4.1–4.3.

There exists a variety of papers dealing with a game theoretic approach to the separable cost or related cost allocation methods, e.g., Driessen and Tijs (1985b); Heaney (1979); Heaney and Dickinson (1982); Legros (1984, 1986); Straffin and Heaney (1981); Tijs and Driessen (1986a). In addition, there is a growing literature on applications of game theory to the cost allocation problem (Young, 1985b).

## 3. The nucleolus in comparison with the egalitarian nonseparable cost method

In the previous section we have established that the egalitarian nonseparable cost (ENSC-) method agrees with the $\tau$-value concept on the class of cost games satisfying the 1-convexity condition (4.6) in terms of costs. Now we weaken the condition (4.6) by requiring that, in case the joint costs c(N) are allocated in such a way that all users outside a nontrivial coalition S pay their separable costs $SC_j(c)$, $j \in N-S$, then the remaining joint costs are at most the cost of the involved coalition S added to a fraction of the nonnegative nonseparable cost NSC(c). This fraction depends upon the coalition size. Formally, the weakened version of the 1-convexity condition (4.6) requires that $NSC(c) \geq 0$ and for all $S \subset N$, $S \neq N,\emptyset$,

$$c(N) \leq c(S) + \sum_{j \in N-S} SC_j(c) + n^{-1}(n-|S|-1)NSC(c). \quad (4.9)$$

Obviously, (4.6) implies (4.9), but the ENSC-method does not any more agree with the $\tau$-value on the enlarged class of cost games satisfying the condition (4.9). Nevertheless, it appears that the ENSC-method on this subclass of cost games can still be regarded as a well-known game theoretic solution concept because it equals the nucleolus concept in a natural way. For that purpose, we first prove that the nucleolus of an associated savings game v is determined by the egalitarian partition of the joint concession amount $g^v(N)$ of the grand coalition, on the understanding that any player i has already received his upper payoff $b^v_i$, $i \in N$.

THEOREM 3.1. Let $v \in G^n$ be the savings game corresponding to the cost game $c \in CG^n$.

(i) Then (4.9) holds for the cost game c if and only if

$$0 \le n^{-1}g^v(N) \le (|S|+1)^{-1}g^v(S) \qquad (4.10)$$
for all $S \subset N$, $S \ne N,\emptyset$.

(ii) If (4.10) holds for the savings game v, then

$$\eta(v) = b^v - n^{-1}g^v(N)1_n \in C(v).$$

PROOF. (i) The equivalence of the two conditions (4.9) and (4.10) is a direct consequence of Lemma 2.3.
(ii) Suppose that (4.10) holds. Put $x := b - n^{-1}g(N)1_n$. By using (1.6) and (3.2), we obtain

$$e(S,x) = v(S) - x(S) = v(S) - b(S) + n^{-1}|S|g(N)$$
$$= -g(S) + n^{-1}|S|g(N) \qquad \text{for all } S \subset N \qquad \text{and hence,}$$
$$e(S,x) \le -n^{-1}g(N) \le 0 \qquad \text{for all } S \subset N,\ S \ne N,\emptyset,$$

where the inequalities follow from (4.10). From this and (2.8) we deduce that $x \in C(v)$. In order to prove $x = \eta(v)$, we apply Theorem II.8.8(i)-(ii). Clearly, $x = \eta(v)$ in case $n = 1$. Thus, let $n \ge 2$. In view of Lemma III.1.2(iii), we have that for all $i \in N$

$$e(N-\{i\},x) = -g(N-\{i\}) + n^{-1}(n-1)g(N) = -n^{-1}g(N).$$

Now we conclude from (2.27)-(2.30) that $\epsilon^1(x) = -n^{-1}g(N)$ and $\{N-\{i\} \mid i \in N\} \subset \mathcal{B}^1(x) \subset \mathcal{B}^j(x)$ for all $1 \le j \le \kappa(x)$. It is sufficient to verify part (ii) of Theorem II.8.8. Let $1 \le j \le \kappa(x)$ and $y \in \mathbb{R}^n$ be such that $y(N) = 0$ and $y(S) \ge 0$ for all $S \in \Sigma_0(x) \cup \mathcal{B}^j(x)$. Then $y(N-\{i\}) \ge 0$ for all $i \in N$. Together with $y(N) = 0$, this implies $y_i = 0$ for all $i \in N$. Particularly, $y(S) = 0$ for all $S \in \mathcal{B}^j(x)$. So, part (ii) of Theorem II.8.8 holds and hence, $x = \eta(v)$. □

THEOREM 3.2. Let $v \in G^n$ be the savings game corresponding to the cost game $c \in CG^n$. In case (4.9) holds for the cost game c, then

$$\eta_i(v) = c(\{i\}) - ENSC_i(c) \qquad \text{for all } i \in N.$$

PROOF. Suppose that the condition (4.9) holds. It follows from Theorem 3.1, Lemma 2.3 and (4.4) respectively that

$$\eta_i(v) = b_i^v - n^{-1}g^v(N) = c(\{i\}) - SC_i(c) - n^{-1}NSC(c)$$

$$= c(\{i\}) - ENSC_i(c) \qquad \text{for all } i \in N. \qquad \square$$

Since (4.6) implies (4.9), we conclude from the Theorems 2.4(i) and 3.2 that the ENSC-method on the class of cost games satisfying the 1-convexity condition (4.6) agrees with the $\tau$-value as well as the nucleolus.
Generally speaking, the semiconvexity condition (4.7) does not guarantee that (4.9) holds. Nevertheless, it appears that (4.9) holds for semiconvex cost games c which satisfy the following condition: for all $i \in N$

$$c(N) \leq c(\{i\}) + \sum_{j \in N-\{i\}} SC_j(c) + n^{-1}NSC(c). \tag{4.11}$$

Obviously, the condition (4.11) for an n-person cost game differs from the condition (4.9) applied to the one-person coalitions in that the fraction of the nonseparable cost is equal to $n^{-1}$ instead of $n^{-1}(n-2)$. The condition (4.11) is equivalent to the requirement

$$n^{-1}(n-1)NSC(c) \leq c(\{i\}) - SC_i(c) \qquad \text{for all } i \in N.$$

According to this reformulation of the condition (4.11) for a cost game c, the part of the alternate cost of any user that can be avoided by including the user in the joint project, is at least a fraction of the nonseparable cost. For cost games with many users, the fraction is almost one.

THEOREM 3.3. Let $v \in G^n$ be the savings game corresponding to the cost game $c \in CG^n$. If both (4.7) and (4.11) hold for the cost game c, then (4.9) holds and

$$\tau_i(v) = c(\{i\}) - ACA_i(c) \qquad \text{for all } i \in N,$$

$$\eta_i(v) = c(\{i\}) - ENSC_i(c) \qquad \text{for all } i \in N.$$

PROOF. In view of the Theorems 2.4(ii) and 3.2, it suffices to show that (4.7) and (4.11) imply (4.9). Suppose that both (4.7) and (4.11) hold. We first prove $NSC(c) \geq 0$. By applying (4.7) to $S = N$, we obtain

$$NSC(c) = c(N) - \sum_{j \in N} SC_j(c) \geq c(\{k\}) - SC_k(c) \geq 0$$

where $k \in N$. Let $S \subset N$ be such that $S \neq N, \emptyset$. Choose $i \in S$. In case $|S| = n-1$, then the equality holds in (4.9) because of (4.2). So, it remains to consider the case $|S| \leq n-2$. Then we have $NSC(c) \leq (n-|S|-1)NSC(c)$. Now it follows from (4.11) and (4.7) respectively that

$$c(N) \leq c(\{i\}) + \sum_{j \in S-\{i\}} SC_j(c) + \sum_{j \in N-S} SC_j(c) + n^{-1}NSC(c)$$

$$\leq c(S) + \sum_{j \in N-S} SC_j(c) + n^{-1}(n-|S|-1)NSC(c).$$

We conclude that the condition (4.9) holds. This completes the proof of the statement mentioned in the theorem. □

EXAMPLE 3.4. Let the cost game $c \in CG^3$ be given by

| coalition S | 1 | 2 | 3 | 12 | 13 | 23 | N |
|---|---|---|---|---|---|---|---|
| cost c(S) | 40 | 50 | 60 | 80 | 90 | 100 | $\alpha$ |
| savings v(S) | 0 | 0 | 0 | 10 | 10 | 10 | $150-\alpha$ |

where $\alpha$ is any real number satisfying $0 \leq \alpha \leq 140$. The associated savings game $v \in G^3$ is listed as well. Then the separable and nonseparable costs in the cost game are given by $SC(c) = (\alpha-100, \alpha-90, \alpha-80)$ and $NSC(c) = 270-2\alpha$. It is straightforward to verify that for the 3-person cost game c

(4.6) holds iff $130 \leq \alpha \leq 135$,
(4.7) holds iff $0 \leq \alpha \leq 130$,
(4.11) holds iff $120 \leq \alpha \leq 140$,
(4.9) holds iff $120 \leq \alpha \leq 135$.

Notice that the symmetry and efficiency properties of the nucleolus and the $\tau$-value yield the egalitarian division of the total savings v(N) by both solution concepts, i.e.,

$$\eta_i(v) = \tau_i(v) = 50-\frac{\alpha}{3} \qquad \text{for all } i \in N. \text{ Consequently,}$$

$$ACA(c) = ENSC(c) = (\frac{\alpha}{3}-10, \frac{\alpha}{3}, \frac{\alpha}{3}+10) \qquad \text{if } 120 \leq \alpha \leq 130$$

which cost allocations follow immediately from Theorem 3.3.

| S | c(S) | v(S) | $g^v(S)$ | $\frac{g^v(S)}{\lvert S\rvert+1}$ | S | c(S) | v(S) | $g^v(S)$ | $\frac{g^v(S)}{\lvert S\rvert+1}$ |
|---|---|---|---|---|---|---|---|---|---|
| ∅ | 0 | 0 | 0 | — | 123 | 440.0 | 49.7 | 306.1 | 76.5 |
| 1 | 0 | 0 | 116.4 | 58.2 | 124 | 518.1 | 229.5 | 247.1 | 61.8 |
| 2 | 0 | 0 | 116.7 | 58.3 | 125 | 490.2 | 259.6 | 280.8 | 70.2 |
| 3 | 489.7 | 0 | 122.7 | 61.3 | 134 | 998.0 | 239.3 | 243.3 | 60.8 |
| 4 | 747.6 | 0 | 243.5 | 121.7 | 135 | 961.5 | 278.0 | 268.4 | 67.1 |
| 5 | 749.8 | 0 | 307.3 | 153.6 | 145 | 1069.2 | 428.2 | 239.0 | 59.7 |
| 12 | 0 | 0 | 233.1 | 77.7 | 234 | 948.2 | 289.1 | 193.8 | 48.4 |
| 13 | 440.0 | 49.7 | 189.4 | 63.1 | 235 | 950.7 | 288.8 | 257.9 | 64.5 |
| 14 | 700.5 | 47.1 | 312.8 | 104.3 | 245 | 1069.2 | 428.2 | 239.3 | 59.8 |
| 15 | 694.0 | 55.8 | 367.9 | 122.6 | 345 | 1554.3 | 432.8 | 240.7 | 60.2 |
| 23 | 486.4 | 3.3 | 236.1 | 78.7 | 1234 | 865.4 | 371.9 | 227.4 | 45.5 |
| 24 | 546.3 | 201.3 | 158.9 | 53.0 | 1235 | 803.8 | 435.7 | 227.4 | 45.5 |
| 25 | 512.2 | 237.6 | 186.4 | 62.1 | 1245 | 940.9 | 556.5 | 227.4 | 45.5 |
| 34 | 1106.5 | 130.8 | 235.4 | 78.5 | 1345 | 1424.6 | 562.5 | 227.4 | 45.5 |
| 35 | 1108.3 | 131.2 | 298.8 | 99.6 | 2345 | 1424.3 | 562.8 | 227.4 | 45.5 |
| 45 | 1209.0 | 288.4 | 262.4 | 87.5 | 12345 | 1307.9 | 679.2 | 227.4 | — |

TABLE 4.4. Cost and saving figures in $10^8$ yen.

| agent $i$ | $ACA_i(c)$ | $ENSC_i(c)$ | $c(i)-\eta_i(v)$ | $c(i)-\tau_i(v)$ | $c(i)-\phi_i(v)$ |
|---|---|---|---|---|---|
| 1 | -87.20 | -70.92 | -70.92 | -78.65 | -71.68 |
| 2 | -87.43 | -71.22 | -71.22 | -78.85 | -100.94 |
| 3 | 397.78 | 412.48 | 412.48 | 406.80 | 395.12 |
| 4 | 565.17 | 549.58 | 549.58 | 555.64 | 554.32 |
| 5 | 519.58 | 487.98 | 487.98 | 502.96 | 531.08 |

TABLE 4.5. Cost allocations by the ACA-, the ENSC-method, the nucleolus $\eta$, the $\tau$-value $\tau$ and the Shapley value $\phi$ (costs in $10^8$ yen).

TABLES 4.4-4.5 are related to the cost allocation problem for the cooperative water resource development in Japan. The two agricultural associations and the three city water service authorities are denoted as 1, 2, 3, 4, and 5 respectively.

We also look at the cooperative water resource development in Japan as described in Section I.7. Particularly, we pay attention to the numerical example in Suzuki and Nakayama (1976) who considered a cost game c with two agricultural associations (denoted as 1 and 2) and three city water service authorities (denoted as 3, 4, and 5). The relevant cost figures as well as the corresponding saving figures are listed in Table 4.4. In the same table we observe that the associated savings game v is neither semiconvex nor 1-convex, but nevertheless, the game v satisfies the allied condition (4.10). Hence, both the nucleolus of the savings game v and the nucleolus cost allocation can be determined with the aid of Theorem 3.1(ii). Suzuki and Nakayama (1976) found the nucleolus cost allocation by solving just one linear programming problem. Table 4.5 compares the cost allocations corresponding to the nucleolus, the $\tau$-value and the Shapley value with the allocations by the ENSC- and the ACA-method. The cost allocations by the nucleolus and the ENSC-method coincide since (4.9) holds for the cost game c. The reader is invited to study the two Tables 4.4–4.5.

## 4. The airport cost allocation problem

Here we treat a game theoretic approach to the cost allocation problem of setting airport landing charges for different types of aircraft. For that purpose, we consider the $\tau$-value as well as the Shapley value cost allocation for the airport cost game (N;c) as defined in Section I.5. In order to determine the $\tau$-value, we first show that the cost game c is semiconvex. We recall that the airport cost game c is given by $c(\emptyset) = 0$ and for all $S \subset N$, $S \neq \emptyset$,

$$c(S) = \max\left(C_j \mid 1 \le j \le m,\ S \cap N_j \neq \emptyset\right) \tag{4.12}$$

where $0 = C_0 < C_1 < C_2 < \cdots < C_m$.

PROPOSITION 4.1. The airport cost game (N;c) of (4.12) satisfies the condition (4.7).

PROOF. By the formula (4.12), we have $c(N) = C_m$ as well as $c(\{i\}) = C_j$ whenever $i \in N_j$, $c(\{i\}) \le c(S)$ whenever $i \in S$. We distinguish the two cases $|N_m| = 1$ and $|N_m| \ge 2$.

(i) Suppose $|N_m| \geq 2$. Then $c(N-\{i\}) = C_m$ for all $i \in N$ and hence, we obtain the result that for all $i \in N$

$$SC_i(c) = c(N) - c(N-\{i\}) = C_m - C_m = 0. \tag{4.13}$$

From this we deduce that the condition (4.7) reduces to the trivial statement $c(\{i\}) \leq c(S)$ whenever $i \in S$.
(ii) Suppose $|N_m| = 1$. Write $N_m = \{\hat{\imath}\}$. By (4.12), we have that $c(N-\{i\}) = C_m$ for all $i \neq \hat{\imath}$, $c(N-\{\hat{\imath}\}) = C_{m-1}$ and thus,

$$SC_i(c) = 0 \quad \text{for all } i \neq \hat{\imath}, \quad SC_{\hat{\imath}}(c) = C_m - C_{m-1}. \tag{4.14}$$

In order to prove the condition (4.7), let $i \in N$ and $S \subset N$ be such that $i \in S$. If $\hat{\imath} \notin S-\{i\}$, then (4.7) reduces to the trivial inequality $c(\{i\}) \leq c(S)$. If $\hat{\imath} \in S-\{i\}$, then we have $c(S) = C_m$ as well as $c(\{i\}) \leq C_{m-1}$ and it follows that

$$c(\{i\}) + \sum_{k \in S-\{i\}} SC_k(c) \leq C_{m-1} + SC_{\hat{\imath}}(c) = C_m = c(S).$$

(iii) From part (i) and (ii) we conclude that (4.7) holds.□

The semiconvexity of the airport cost game is equivalent to the semiconvexity of the associated savings game. Hence, the $\tau$-value $\tau(v)$ of the savings game $v$ can be determined with the aid of Theorem III.6.2. Subsequently, the $\tau$-value cost allocation $\tau(c)$ for the airport game $c$ is obtained by $\tau_i(c) = c(\{i\}) - \tau_i(v)$ for all $i \in N$. Its exact expression in terms of the runway costs $C_j$, $1 \leq j \leq m$, is formulated in the next theorem.

THEOREM 4.2. Let $(N;c)$ be the airport cost game of (4.12) and put $n_j := |N_j|$ for all $1 \leq j \leq m$. Then

(i) $\tau_i(c) = \left(\sum_{k=1}^{m} n_k C_k\right)^{-1} C_m C_j$ for $i \in N_j$ in case $n_m \geq 2$.

(ii) If $n_m = 1$ and $m \geq 2$, then

$$\tau_i(c) = \left(\sum_{k=1}^{m-1} n_k C_k + C_{m-1}\right)^{-1} C_{m-1} C_j \qquad \text{for } i \in N_j,\ j \neq m,$$

$$\tau_{\hat{\imath}}(c) = \tau_i(c) + C_m - C_{m-1} \qquad \text{for } \hat{\imath} \in N_m \text{ and } i \in N_{m-1}.$$

PROOF. Let $(N;v)$ be the associated savings game. As such, $v$ is a semiconvex savings game because of Proposition 4.1. Further, the savings game $v$ is zero-normalized and $I(v) \neq \emptyset$ due to the superadditivity of $v$.
By (4.1), $c(N) = C_m$ and $c(\{i\}) = C_j$ whenever $i \in N_j$, we get

$v(N) = \sum_{k=1}^{m} n_k C_k - C_m$. Note that $v(N) = \sum_{k=1}^{m-1} n_k C_k$ if $n_m = 1$.

(i) Let $n_m \geq 2$. By (4.13), we have $SC_i(c) = 0$ for all $i \in N$ and hence, $NSC(c) = c(N) = C_m > 0$. By Lemma 2.3, we obtain $g^v(N) = NSC(c) > 0$, $b_i^v = c(\{i\}) - SC_i(c) = C_j$ for $i \in N_j$. Now it follows from Theorem III.6.2(iii) that for $i \in N_j$

$$\begin{aligned}\tau_i(c) &= c(\{i\}) - \tau_i(v) = c(\{i\}) - v(N)\left(b^v(N)\right)^{-1} b_i^v \\ &= C_j - v(N)\left(b^v(N)\right)^{-1} C_j = \left(b^v(N) - v(N)\right)\left(b^v(N)\right)^{-1} C_j \\ &= g^v(N)\left(b^v(N)\right)^{-1} C_j = \left(\sum_{k=1}^{m} n_k C_k\right)^{-1} C_m C_j .\end{aligned}$$

(ii) Let $n_m = 1$ and $m \geq 2$. Write $N_m = \{\hat{\imath}\}$. By (4.14), we have $SC_i(c) = 0$ for all $i \neq \hat{\imath}$, $SC_{\hat{\imath}}(c) = C_m - C_{m-1}$ and thus, $NSC(c) = c(N) - SC_{\hat{\imath}}(c) = C_{m-1} > 0$. By Lemma 2.3, we obtain $g^v(N) = NSC(c) > 0$, $b_i^v = C_j$ for $i \in N_j$, $i \neq \hat{\imath}$, $b_{\hat{\imath}}^v = C_{m-1}$. From this and Theorem III.6.2(iii), it is straightforward to deduce the formula for the $\tau$-value cost allocation $\tau(c)$ as given in the statement (ii). □

By the above theorem, the $\tau$-value concept on the class of airport cost games agrees with the following simple cost allocation rule.
"Whenever there are at least two landings by planes of the largest type, allocate the joint costs $C_m$ in proportion to the runway costs $C_j$, $1 \leq j \leq m$, where cost $C_j$ represents the cost of a runway adequate for planes of type $j$. In case the planes of the largest type use the runway only once, as an initial step, charge the incremental cost $C_m - C_{m-1}$ to the largest plane and in the second step, allocate the

remaining joint costs $C_{m-1}$ among all landings in proportion to the runway costs of their type, on the understanding that the largest plane of type m is now regarded as a plane of type m-1."

Alternatively, the allocation of the total landing fees can be based on the Shapley value. A simple and exact expression for the Shapley value cost allocation in terms of the runway costs is due to Littlechild and Owen (1973).

THEOREM 4.3. Let (N;c) be the airport cost game of (4.12) and put $m_j := \sum_{k=j}^{m} |N_k|$ for all $1 \le j \le m$. Then

$$\phi_i(c) = \sum_{k=1}^{j} m_k^{-1}(C_k - C_{k-1}) \qquad \text{for } i \in N_j.$$

PROOF. The essential idea in the proof is to write the cost game of (4.12) as the sum of m related cost games for which the Shapley values can easily be determined with the aid of the standard properties for values. Formally, for any type $j = 1, 2, \cdots, m$, we define the cost game $(N;c_j)$ by

$$\begin{aligned} c_j(S) &:= C_j - C_{j-1} && \text{if } S \cap M_j \neq \emptyset \\ &= 0 && \text{if } S \cap M_j = \emptyset \end{aligned}$$

where $M_j := \bigcup_{k=j}^{m} N_k$ represents the set of all landings by planes of type j and larger planes.

(i) We assert that $c(S) = \sum_{k=1}^{m} c_k(S)$ for all $S \subset N$. (4.15)

Let $S \subset N$, $S \neq \emptyset$. By the definition of the airport game c, there exists a unique $1 \le j \le m$ such that $c(S) = C_j$. Then we obtain $S \cap N_j \neq \emptyset$ and $S \cap N_k = \emptyset$ for all $j < k \le m$. Particularly, $S \cap M_k = \emptyset$ iff $j < k \le m$. Now it follows that

$$\sum_{k=1}^{m} c_k(S) = \sum_{k=1}^{j} (C_k - C_{k-1}) = C_j - C_0 = C_j = c(S).$$

So, (4.15) holds.

(ii) Let $1 \le j \le m$. We assert that any $i \in N-M_j$ is a dummy in the game $c_j$. Let $i \in N-M_j$ and $S \subset N-\{i\}$. Then we have $c_j(\{i\}) = 0$ and further, $S \cap M_j \neq \emptyset$ iff $(S \cup \{i\}) \cap M_j \neq \emptyset$. It follows that $c_j(S \cup \{i\}) - c_j(S) = 0 = c_j(\{i\})$ and thus, any $i \in N-M_j$ is a dummy in the game $c_j$. Now we conclude from the dummy player property of the Shapley value that

$$\phi_i(c_j) = c_j(\{i\}) = 0 \qquad \text{for all } i \in N-M_j.$$

(iii) Let $1 \le j \le m$. We assert that the players in $M_j$ are symmetric in the game $c_j$. Let $\theta\colon N \to N$ be a permutation and let $i_1 \in M_j$, $i_2 \in M_j$ be such that $\theta(i_1) = i_2$, $\theta(i_2) = i_1$ and $\theta(i) = i$ for all $i \in N-\{i_1,i_2\}$. For any coalition $S \neq \emptyset$ we have the equivalence $S \cap M_j \neq \emptyset$ iff $\theta S \cap M_j \neq \emptyset$ and so, $(\theta c_j)(\theta S) = c_j(S) = c_j(\theta S)$. It follows that $\theta c_j = c_j$. Now we conclude from the symmetry property of the Shapley value that

$$\phi_{i_1}(c_j) = \phi_{\theta(i_2)}(\theta c_j) = \phi_{i_2}(c_j) \qquad \text{for all } i_1, i_2 \in M_j.$$

(iv) Let $1 \le j \le m$. Together with the parts (ii)-(iii), the efficiency property of the Shapley value yields

$$\begin{aligned} \phi_i(c_j) &= 0 && \text{if } i \in N-M_j \\ &= m_j^{-1}(C_j - C_{j-1}) && \text{if } i \in M_j. \end{aligned}$$

From this, the formula (4.15) and the additivity property of the Shapley value, it follows that for all $i \in N_j$

$$\phi_i(c) = \phi_i\Big(\sum_{k=1}^{m} c_k\Big) = \sum_{k=1}^{m} \phi_i(c_k) = \sum_{k=1}^{j} m_k^{-1}(C_k - C_{k-1}). \square$$

By Theorem 4.3, the Shapley value concept on the class of airport cost games agrees with the following simple cost allocation rule proposed in both Baker (1965) and Thompson (1971) concerning their research on the determination of airport landing charges for different types of aircraft.

"Divide the cost $C_1$ of a runway adequate for planes of the smallest type equally among the number of landings by all planes. Divide the incremental cost $C_2 - C_1$ of a runway adequate to receive all landings by planes of the second smallest type (above the cost of the smallest type) equally among the number of landings by all but the smallest type. Continue thus until finally the incremental cost $C_m - C_{m-1}$ of the largest type (above the cost of the second largest type) is divided equally among the number of landings made by the largest aircraft type."
For an axiomatization of the Shapley value on an enlarged class of airport cost games, we refer to Dubey (1982).

The determination of the nucleolus cost allocation for the airport game is rather technical and there is no simple formula for the nucleolus cost allocation available. Here we omit the treatment of the nucleolus cost allocation for which we refer to Owen (1982, pp. 259-261).

As a numerical example of an airport game, we consider the total landings at the Birmingham airport (Birmingham, U.K.) during the year 1968-1969. The data is taken from Thompson (1971) and is listed in Table 4.6. In particular, there are 13,572 landings of 11 different aircraft types. The landing fee for a plane of type j is composed of an operating charge $a_j$ on a per-landing basis as well as a capital charge $\psi_j$ which has been computed according to the Shapley value, the nucleolus and the $\tau$-value respectively. The reader is invited to study Table 4.6 and to compare the resulting landing fees $\psi_j + a_j$ with the actual landing fees $f_j$, $1 \le j \le 11$.

At the beginning of the section we have shown that an airport cost game generates a semiconvex savings game. As we shall see in the next chapter, the associated savings game satisfies a much more significant condition which is known as the convexity condition for a savings game.

| Aircraft type | $j$ | $C_j$ | $n_j$ | $a_j$ |
|---|---|---|---|---|
| Fokker Friendship 27 | 1 | £ 65,899 | 42 | £ 5.23 |
| Viscount 800 | 2 | 76,725 | 9,555 | 6.09 |
| Hawker Siddeley Trident | 3 | 95,200 | 288 | 7.55 |
| Britannia 100 | 4 | 97,200 | 303 | 7.71 |
| Caravelle VLR | 5 | 97,436 | 151 | 7.73 |
| BAC 111 (500) | 6 | 98,142 | 1,315 | 7.79 |
| Vanguard 953 | 7 | 102,496 | 505 | 8.13 |
| Comet 4B | 8 | 104,849 | 1,128 | 8.32 |
| Britannia 300 | 9 | 113,322 | 151 | 8.99 |
| Convair Corronado | 10 | 115,440 | 112 | 9.16 |
| Boeing 707 | 11 | 117,676 | 22 | 9.34 |

| $j$ | $\phi_j$ | $\phi_j + a_j$ | $\eta_j$ | $\eta_j + a_j$ | $\tau_j$ | $\tau_j + a_j$ | $f_j$ |
|---|---|---|---|---|---|---|---|
| 1 | £ 4.86 | £ 10.09 | £ 7.89 | £ 13.12 | £ 6.81 | £ 12.04 | £ 5.80 |
| 2 | 5.66 | 11.75 | 7.89 | 13.98 | 7.93 | 14.02 | 11.40 |
| 3 | 10.30 | 17.85 | 7.89 | 15.44 | 9.83 | 17.38 | 21.70 |
| 4 | 10.85 | 18.56 | 7.89 | 15.60 | 10.04 | 17.75 | 29.80 |
| 5 | 10.92 | 18.65 | 7.89 | 15.62 | 10.06 | 17.79 | 20.30 |
| 6 | 11.13 | 18.92 | 7.89 | 15.68 | 10.14 | 17.93 | 16.70 |
| 7 | 13.40 | 21.53 | 7.89 | 16.02 | 10.59 | 18.72 | 26.40 |
| 8 | 15.07 | 23.39 | 7.89 | 16.21 | 10.83 | 19.15 | 29.40 |
| 9 | 44.80 | 53.79 | 40.16 | 49.15 | 11.71 | 20.70 | 34.70 |
| 10 | 60.61 | 69.77 | 40.16 | 49.32 | 11.92 | 21.08 | 48.30 |
| 11 | 162.24 | 171.58 | 103.46 | 112.80 | 12.16 | 21.50 | 66.70 |

TABLE 4.6. Aircraft landings, runway costs and landing fees at Birmingham Airport during the year 1968–1969. Here $n_j$ denotes the number of aircraft landings by planes of type j and

$C_j$ the annual capital cost,

$a_j$ the operating cost per landing,

$f_j$ the actual landing fee

for a plane of type j, $1 \le j \le 11$. Further, $\psi_j$ denotes the capital charge for a plane of type j according to the value $\psi$ where $\psi \in \{\phi,\eta,\tau\}$. So, $\psi_j = \psi_i(c)$ whenever $i \in N_j$.

CHAPTER V

# CONVEX GAMES AND SOLUTION CONCEPTS

## 1. Convex games

In analysis, convexity is a very natural property of various real-valued functions and their domains. A function $f\colon X \to \mathbb{R}$ is called convex on its convex domain $X$ if for all $x \in X$, $y \in X$ and $\alpha \in [0,1]$

$$f\big(\alpha x + (1-\alpha)y\big) \le \alpha f(x) + (1-\alpha)f(y). \tag{5.1}$$

From the angle of geometry, the convexity condition (5.1) for the function $f\colon X \to \mathbb{R}$ requires that the straight line segment with end points $\big(x, f(x)\big)$ and $\big(y, f(y)\big)$ lies above or on the graph of the real-valued function $f$.
As examples of convex functions on the linear space $\mathbb{R}^n$, we mention the components of the maximum excess map. For any n-person game $v$ and any player $i$, the i-th component function $ME_i^v\colon \mathbb{R}^n \to \mathbb{R}$ of the corresponding maximum excess map $ME^v\colon \mathbb{R}^n \to \mathbb{R}^n$ is given by the formula (3.8) and it is straightforward to verify the convexity of the function $ME_i^v$ on $\mathbb{R}^n$, i.e., for all $x \in \mathbb{R}^n$, $y \in \mathbb{R}^n$ and $\alpha \in [0,1]$

$$ME_i^v\big(\alpha x + (1-\alpha)y\big) \le \alpha ME_i^v(x) + (1-\alpha)ME_i^v(y).$$

From elementary calculus, it is well-known that a twice differentiable function $f\colon I \to \mathbb{R}$ on an open interval $I \subset \mathbb{R}$ is convex if and only if

its first derivative $f'$ is nondecreasing
(i.e., $f'(x) \ge f'(y)$ whenever $x \ge y$) or equivalently,

its second derivative $f''$ is nonnegative
(i.e., $f''(x) \ge 0$ for all $x \in I$).

These two characterizations of convexity are samples of the mathematically beautiful and practically useful theory on convexity developed in advanced calculus and other fields. Many mathematical models in economics are described with the aid of production functions which are said to be convex because of the "increasing returns to scale". The strictly convex production function f in the production economy with landowners and peasants as will be presented in Example 2.2, is a case in point. The involved production economy with m landless peasants and only one landowner generates the (m+1)-person game (N;v) of Section I.3. Here it appears that the real-valued set-function $v: 2^N \to \mathbb{R}$ is a so-called supermodular function whenever the production function f is strictly convex. A real-valued set-function is said to be <u>supermodular</u> if it satisfies one of the following seven equivalent conditions.

PROPOSITION 1.1. Let N be a finite nonempty set and let $F_N := \{w \mid w: 2^N \to \mathbb{R}\}$. For any coalition $T \subset N$, the corresponding difference operator $\Delta_T: F_N \to F_N$ is defined by

$$(\Delta_T w)(S) := w(S \cup T) - w(S-T) \quad \text{for all } w \in F_N,\ S \subset N.$$

The following statements for $v: 2^N \to \mathbb{R}$ are equivalent.

(i) $v(S \cup \{i\}) - v(S) \leq v(T \cup \{i\}) - v(T)$ (5.2)
for all $i \in N$ and all $S, T \subset N$
such that $S \subset T \subset N-\{i\}$.

(ii) $g^v(S \cup \{i\}) - g^v(S) \geq g^v(T \cup \{i\}) - g^v(T)$ (5.3)
for all $i \in N$ and all $S, T \subset N$
such that $S \subset T \subset N-\{i\}$.

(iii) $v(S) + v(T) \leq v(S \cup T) + v(S \cap T)$ (5.4)
for all $S, T \subset N$.

(iv) $g^v(S) + g^v(T) \geq g^v(S \cup T) + g^v(S \cap T)$ (5.5)
for all $S, T \subset N$.

(v) $v(S \cup R) - v(S) \leq v(T \cup R) - v(T)$ (5.6)
for all $R, S, T \subset N$ such that $S \subset T \subset N-R$.

(vi) $g^v(S \cup R) - g^v(S) \geq g^v(T \cup R) - g^v(T)$ (5.7)
for all $R, S, T \subset N$ such that $S \subset T \subset N-R$.

(vii) $(\Delta_S(\Delta_T v))(R) \geq 0$ for all $R, S, T \subset N$. (5.8)

PROOF. By (3.2), it is obvious that (5.3), (5.5) and (5.7) respectively are reformulations of (5.2), (5.4) and (5.6) in terms of the corresponding gap function. Therefore, it is sufficient to prove (i) ⇒ (iii) ⇒ (v) ⇒ (vii) ⇒ (i).

(a) Suppose that (i) holds. Let $S \subset N$ and $T \subset N$. Clearly, the equality holds in (5.4) whenever $S \subset T$. It remains to consider the case $S-T \neq \emptyset$. Write $S-T = \{i_1, i_2, \ldots, i_k\}$ where $k := |S-T| \geq 1$. Put $R := S \cap T$, $[i_0] := \emptyset$ and in addition, $[i_j] := \{i_1, i_2, \ldots, i_j\}$ for all $1 \leq j \leq k$. By (5.2), we get

$$v(R \cup [i_j]) - v(R \cup [i_{j-1}]) \leq v(T \cup [i_j]) - v(T \cup [i_{j-1}])$$

for all $1 \leq j \leq k$. Now it follows that

$$\begin{aligned} v(S) - v(S \cap T) &= v(R \cup [i_k]) - v(R) \\ &= \sum_{j=1}^{k} \left(v(R \cup [i_j]) - v(R \cup [i_{j-1}])\right) \\ &\leq \sum_{j=1}^{k} \left(v(T \cup [i_j]) - v(T \cup [i_{j-1}])\right) \\ &= v(T \cup [i_k]) - v(T) = v(S \cup T) - v(T). \end{aligned}$$

Thus, (5.4) holds whenever $S-T \neq \emptyset$. So, (i) implies (iii).
(b) Suppose that (iii) holds. Let $R, S, T \subset N$ be such that $S \subset T \subset N-R$. Now it follows immediately from (5.4) that

$$v(S \cup R) + v(T) \leq v(T \cup R) + v(S).$$

Thus, (5.6) holds. So, (iii) implies (v).
(c) Suppose that (v) holds. Let $R, S, T \subset N$. The relevant condition $\left(\Delta_S(\Delta_T v)\right)(R) \geq 0$ is equivalent to the inequality

$$v\left(R \cup S \cup T\right) - v\left((R \cup S)-T\right) \geq v\left((R-S) \cup T\right) - v\left((R-S)-T\right)$$

which holds because of (5.6). So, (v) implies (vii).
(d) Suppose that (vii) holds. Let $i \in N$, $S \subset N$ and $T \subset N$ be such that $S \subset T \subset N-\{i\}$. In view of the condition (5.8), we have $\left(\Delta_{T-S}(\Delta_{\{i\}} v)\right)(S) \geq 0$ which inequality is equivalent to

$$v(T \cup \{i\}) - v(T) \geq v(S \cup \{i\}) - v(S).$$

Thus, (5.2) holds. So, (vii) implies (i). □

Nowadays the term supermodularity is more popular than the equivalent term convexity with respect to real-valued set-functions. Nevertheless, a game $(N;v)$ is said to be <u>convex</u> if its characteristic function $v\colon 2^N \to \mathbb{R}$ is supermodular.

Convex games were introduced in Shapley (1971) who already mentioned the equivalent convexity conditions of the above proposition with the exception of the three conditions (5.3), (5.5) and (5.7) in terms of the corresponding gap function. The term convexity with respect to games is justifiable because the nonnegative "second differences" of the difference operator $\Delta_T$ (see (5.8)) are analogous to the nonnegative second derivative associated with a convex function in real analysis. Notice that the expression $(\Delta_T v)(S)$ represents the difference between the worths of coalition S with and without the players in coalition T. The convexity condition (5.6) expresses that the incentives of an arbitrary fixed coalition for joining another disjoint coalition increase as the coalition grows: the so-called "snowballing" or "bandwagon" effect. In the game theoretic context we prefer the convexity condition (5.2) which expresses the "snowballing" effect with respect to single players. In other words, a game is convex if and only if its characteristic function satisfies nondecreasing marginal returns with respect to the coalition size (or equivalently, its gap function satisfies nonincreasing marginal returns with respect to the coalition size).

Examples of convex games will be given in the next section. In view of (5.2), the convexity notion for a game can be described as the intersection of a finite number of weak inequalities that are linear in the worths of the coalitions in the game. Therefore, the class $C^n$ of convex n-person games is a polyhedral cone in the linear space $G^n$. The class $C^n$ is a full-dimensional cone in $G^n$.

THEOREM 1.2. $\dim C^n = 2^n - 1$.

PROOF. We establish that the set $\{u_T \in G^n \mid T \subset N, T \neq \emptyset\}$ of all unanimity n-person games forms a basis of the class $C^n$. We first show the convexity of the unanimity games by verifying the convexity condition (5.2). Let $T \subset N$, $T \neq \emptyset$. From (1.5) we deduce that for all $i \in N$ and all $S \subset N-\{i\}$

$$\begin{aligned} u_T(S \cup \{i\}) - u_T(S) &= 1 \qquad \text{whenever } i \in T,\ T-\{i\} \subset S \\ &= 0 \qquad \text{otherwise.} \end{aligned}$$

Consequently, $u_T(S_1 \cup \{i\}) - u_T(S_1) \le u_T(S_2 \cup \{i\}) - u_T(S_2)$ for all $i \in N$ and all $S_1, S_2 \subset N$ such that $S_1 \subset S_2 \subset N-\{i\}$.

Hence, the unanimity n-person game $u_T$ is indeed convex. So, it remains to show that the set $\{u_T \in G^n \mid T \subset N, T \neq \emptyset\}$ is linearly independent in order to conclude that this set forms a basis of $C^n$.
Suppose there exist real numbers $\alpha_T$, $T \subset N$, $T \neq \emptyset$, such that $\sum_{\emptyset \neq T \subset N} \alpha_T u_T(S) = 0$ for all $S \subset N$. We prove by induction on $|T|$ that $\alpha_T = 0$ for all $T \subset N$, $T \neq \emptyset$. In case $|T| = 1$, then the equivalence $u_R(T) = 1$ iff $R = T$ holds and hence,

$$0 = \sum_{\emptyset \neq R \subset N} \alpha_R u_R(T) = \alpha_T u_T(T) = \alpha_T.$$

This completes the proof of the induction basis for $|T| = 1$. Let $2 \leq |T| \leq n$ and suppose that $\alpha_R = 0$ for all $R \subset N$ with $1 \leq |R| < |T|$. For all $R \subset N$ with $|R| \geq |T|$ we have the equivalence $u_R(T) = 1$ iff $R = T$. It follows that

$$0 = \sum_{\emptyset \neq R \subset N} \alpha_R u_R(T) = \sum_{R;|R| \geq |T|} \alpha_R u_R(T) = \alpha_T u_T(T) = \alpha_T.$$

Now we conclude that $\alpha_T = 0$ for all $T \subset N$, $T \neq \emptyset$. This completes the inductive proof of the linear independence of the set of all unanimity games with player set N and as such, this set forms a basis of $G^n$ as well as $C^n$. □

Finally, we treat the relationship of the convexity to the term semiconvexity which was introduced in Section III.6. The next proposition states that convexity implies semiconvexity because the gap function of any convex game is monotonic.

PROPOSITION 1.3.

(i) If $v \in C^n$, then $g^v(S) \leq g^v(S \cup \{i\})$ for all $i \in N$ and all $S \subset N$ such that $S \subset N-\{i\}$.

(ii) In particular, $C^n \subset SC^n$.

PROOF. Let $v \in C^n$, $i \in N$ and $S \subset N-\{i\}$. By (5.2), we have $v(S \cup \{i\}) - v(S) \leq v(N) - v(N-\{i\})$. In view of (3.1) and (3.2), this inequality is equivalent to $g(S) \leq g(S \cup \{i\})$. Furthermore, it follows immediately from part (i) that $0 = g(\emptyset) \leq g(\{i\}) \leq g(T)$ for all $i \in N$ and all $T \subset N$ with $i \in T$. We conclude that $v \in C^n$ implies $v \in SC^n$. □

It is obvious from (3.1) and (3.14) that the class $SC^n$ of semiconvex n-person games is a polyhedral cone in the linear space $G^n$. The class $SC^n$ is a full-dimensional cone in $G^n$ because it contains the $(2^n-1)$-dimensional cone $C^n$.

COROLLARY 1.4. $\dim SC^n = 2^n - 1$.

From the convexity condition (5.4) we derive at once that convexity implies superadditivity and zero-monotonicity. Together with the inclusion $C^n \subset SC^n$, this yields that the $\tau$-value of a convex game can usually be studied by means of the Theorems III.6.2–6.3. We note that $C^n = SC^n$ when $n \le 2$, while the inclusion $C^n \subset SC^n$ is strict in case $n \ge 3$. For instance, the semiconvex 4-person game v of Example III.3.7 is not convex since it is not zero-monotonic.

## 2. Examples of convex games

EXAMPLE 2.1. The quadratic function $f: \mathbb{R} \to \mathbb{R}$ defined by $f(x) := x^2$ for all $x \in \mathbb{R}$ is convex due to its nonnegative second derivative $f''$ (i.e., $f''(x) = 2 \ge 0$ for all $x \in \mathbb{R}$). The game theoretic version of the quadratic function on $\mathbb{R}$ is the characteristic function $v: 2^N \to \mathbb{R}$ given by

$$v(S) := \Big(\sum_{j\in S} x_j\Big)^2 = \big(x(S)\big)^2 \qquad \text{for all } S \subset N$$

where the player set $N := \{1,2,\ldots,n\}$ and $x \in \mathbb{R}^n$ is any n-tuple satisfying $x_i \ge 0$ for all $i \in N$. In order to prove that the quadratic set-function v is supermodular, we check the convexity condition (5.2). Let $i \in N$, $S \subset N$ and $T \subset N$ be such that $S \subset T \subset N-\{i\}$. The condition (5.2) requires

$$\big(x(S \cup \{i\})\big)^2 - \big(x(S)\big)^2 \le \big(x(T \cup \{i\})\big)^2 - \big(x(T)\big)^2$$

or equivalently, $x_i\big(x(T-S)\big) \ge 0$ which weak inequality holds because of the nonnegativity of the n-tuple x. Thus, (5.2) holds and hence, the game $(N;v)$ is convex.

EXAMPLE 2.2. We consider again the production economy with landowners and peasants as described in Section I.3. The involved production function $f: \{0,1,\ldots,m\} \to \mathbb{R}$ is said to be strictly convex if the marginal returns $f(t+1) - f(t)$, $0 \le t \le m-1$, of the function form an increasing sequence, i.e., $f(t+1) - f(t) > f(t) - f(t-1)$ for all $1 \le t \le m-1$.

The economic situation with m landless peasants and only one landowner generates the (m+1)-person game (N;v) where its characteristic function v is given by

$$\begin{aligned} v(S) &= 0 && \text{if } 1 \notin S \\ &= f(|S|-1) && \text{if } 1 \in S. \end{aligned}$$

Now we assert that the associated set-function v: $2^N \to \mathbb{R}$ is supermodular whenever the production function f is supposed to be strictly convex. For that purpose, we verify the convexity condition (5.2).
Let $i \in N$, $S \subset N$ and $T \subset N$ be such that $S \subset T \subset N-\{i\}$. Put $s := |S|$ and $t := |T|$. Clearly, the equality holds in (5.2) whenever $1 \notin T$, $i \neq 1$, while (5.2) reduces to

$$\begin{aligned} &f(s) \le f(t) && \text{if } i = 1, \\ &f(t-1) \le f(t) && \text{if } 1 \in T-S, \\ &f(s) - f(s-1) \le f(t) - f(t-1) && \text{if } 1 \in S. \end{aligned}$$

The first two inequalities hold because any production function f is required to be nondecreasing, whereas the last inequality is due to the strict convexity of f. Thus, (5.2) holds and hence, the game (N;v) is convex whenever the production function f is strictly convex.
For the economic situation containing two landowners and concerning a strictly convex production function f as well, the associated (m+2)-person game (M;w) is given by

$$\begin{aligned} w(S) &= 0 && \text{if } S \subset \{3,4,\ldots,m+2\} \\ &= f(|S|-1) && \text{if } S \in \Gamma_{12} \cup \Gamma_{21} \\ &= f(|S|-2) && \text{if } \{1,2\} \subset S. \end{aligned}$$

The convexity condition for the game (M;w) requires

$$w(M-\{2\}) - w(M-\{1,2\}) \le w(M) - w(M-\{1\})$$

which inequality is equivalent to $f(m) \le 0$. In view of the nonnegativity of any production function, we conclude that the game (M;w) where $m \ge 2$ is never convex in spite of the strict convexity of the production function f.

EXAMPLE 2.3. We consider again the bankruptcy problem as presented in Section I.6. We assert that the bankruptcy game $(N;v_{E;d})$ arising from a bankruptcy problem (E;d) is convex. Put $w := v_{E;d}$ and $\alpha := E - d(N)$. In view of (1.2), the characteristic function w is then given by

$$w(S) = \max\big(0, \alpha + d(S)\big) \qquad \text{for all } S \subset N.$$

In order to prove the supermodularity of the set-function $w$, we verify the convexity condition (5.2) for $w$. Let $i \in N$, $S \subset N$ and $T \subset N$ be such that $S \subset T \subset N-\{i\}$. We first note that $\max(0, \beta) + \max(0, \gamma) = \max(0, \beta, \gamma, \beta + \gamma)$ for all $\beta, \gamma \in \mathbb{R}$. From this we deduce

$$w(S \cup \{i\}) + w(T) = \max(0, \alpha + d_i + d(S)) + \max(0, \alpha + d(T))$$

$$= \max(0, \alpha + d_i + d(S), \alpha + d(T), 2\alpha + d_i + d(S) + d(T)).$$

Further, in the same way we obtain that $w(T \cup \{i\}) + w(S)$

$$= \max(0, \alpha + d_i + d(T), \alpha + d(S), 2\alpha + d_i + d(T) + d(S)).$$

We also have $d_i \geq 0$ and $d(S) \leq d(T)$ because $d_j \geq 0$ for all $j \in N$ as well as $S \subset T$. Now it follows immediately that $w(S \cup \{i\}) + w(T) \leq w(T \cup \{i\}) + w(S)$. Thus, (5.2) holds for $w$ and hence, the bankruptcy game $(N;v_{E;d})$ is convex.

EXAMPLE 2.4. We consider again the airport cost allocation problem as described in Section I.5. The characteristic cost function $c$ of the airport cost game $(N;c)$ is given by $c(\emptyset) = 0$ and for all $S \subset N$, $S \neq \emptyset$,

$$c(S) = \max(C_j \mid 1 \leq j \leq m, S \cap N_j \neq \emptyset)$$

where $0 = C_0 < C_1 < C_2 < \cdots < C_m$. Evidently, we have that $c(S) \leq c(T)$ whenever $S \subset T \subset N$. In point of fact, the cost function $c$ is a so-called <u>submodular</u> set-function because it satisfies the following condition: for all $S, T \subset N$

$$c(S) + c(T) \geq c(S \cup T) + c(S \cap T). \tag{5.9}$$

Clearly, the equality holds in (5.9) when $S = \emptyset$ or $T = \emptyset$. In order to prove (5.9), it remains to look at the case $S \neq \emptyset$, $T \neq \emptyset$. Let $1 \leq j \leq m$ be such that $(S \cup T) \cap N_j \neq \emptyset$ and $c(S \cup T) = C_j$ as well. Without loss of generality, we may suppose $S \cap N_j \neq \emptyset$ (otherwise $T \cap N_j \neq \emptyset$, etc.) and so, we get $c(S) \geq C_j$. Now it follows that

$$c(S) + c(T) \geq C_j + c(T)$$
$$\geq C_j + c(S \cap T) = c(S \cup T) + c(S \cap T).$$

Thus, (5.9) holds. In view of both (5.9) and (5.4), it is obvious that the submodularity of the cost function $c$ is equivalent to the supermodularity of the corresponding savings function v given by the formula (4.1). We conclude that the savings game (N;v) associated with the airport cost game (N;c) is convex.

We notice that the cost function c of the numerical example in Suzuki and Nakayama (1976), listed in Table 4.4, is not a submodular set-function because

$$1898.9 = c(234) + c(235) < c(2345) + c(23) = 1910.7.$$

In addition, we note that the characteristic function v of the 3-person game of (2.10) is not supermodular in case $0.5 < \alpha \leq 1$ since $2\alpha = v(12) + v(13) > v(123) + v(1) = 1$ whenever $0.5 < \alpha \leq 1$.

Finally, we investigate the convexity of simple games. In general, a simple game does not satisfy the convexity condition (5.2). For instance, (5.2) is not valid for the simple n-person game v of (1.4) whenever $2 \leq m < n$ because

$$1 = v(T \cup \{2\}) - v(T) > v(N) - v(N-\{2\}) = 0$$

for any coalition $T \subset N-\{2\}$ such that $|T| = m-1$ and $1 \in T$. Notice that $v(J^v) = 0$ whenever $2 \leq m < n$. According to the following theorem, the convexity of a simple game can be characterized by the fact that the set of all veto players is a powerful coalition in the game.

THEOREM 2.5. Let $v \in S^n$. Then $v \in C^n$ iff $v(J^v) = 1$.

PROOF. (i) From both $v \in S^n$, (3.1) and (3.2) we derive that

$$b_i = 1 \text{ if } i \in J, \qquad b_i = 0 \text{ if } i \in N-J \qquad \text{and hence,}$$

$$g(N) = |J| - 1, \qquad g(J) = |J| - v(J).$$

If $v \in C^n$, then it follows from Proposition 1.3(i) that $g(J) \leq g(N)$ which inequality is equivalent to $v(J) \geq 1$. Because $v(J) \in \{0,1\}$, we get that $v(J) = 1$ whenever $v \in C^n$. (ii) To prove the converse statement, suppose $v(J) = 1$. We verify the convexity condition (5.4). Let $S \subset N$ and $T \subset N$. If $v(S) = 0$ or $v(T) = 0$, then (5.4) holds because of the monotonicity of the game v. It remains to consider the case $v(S) = v(T) = 1$. Now we conclude from Proposition III.7.1 that $J \subset S$ as well as $J \subset T$ and so, $J \subset S \cap T \subset S \cup T$. From this, $v(J) = 1$ and the monotonicity of v, we deduce that $v(S \cap T) = v(S \cup T) = 1$. Thus, (5.4) holds and hence, v is a convex game. □

The above theorem states that a simple game is convex if and only if the veto players in the game form a winning coalition. In particular, the convexity of a simple game implies the existence of veto players which is a sufficient condition for the nonemptiness of the core of the simple game by Corollary III.7.4. We conclude that the core of a convex simple game is nonempty. The next section is devoted to the study of the core of an arbitrary convex game.

## 3. The core of a convex game

The purpose of this section is to show that any convex game possesses a nonempty core by describing the structure of the core in terms of so-called marginal worth vectors. We first treat two numerical examples of convex games in order to illustrate the structure of their cores.

EXAMPLE 3.1. Let the 3-person game v be given by

$$v(i) = 0 \quad \text{for all } i \in N, \qquad v(N) = 22,$$
$$v(12) = 4, \qquad v(13) = 6, \qquad v(23) = 12.$$

The game v can be interpreted as the bankruptcy game of (1.2) arising from the bankruptcy problem with the estate $E = 22$ and the three claims $d_1 = 10$, $d_2 = 16$, $d_3 = 18$. Furthermore, the game v can be seen as the zero-normalized 3-person version of the game of Example 2.1 with respect to the 3-tuple $(1,2,3) \in \mathbb{R}^3$ of real numbers, i.e.,

$$v(S) = \Big(\sum_{j\in S} j\Big)^2 - \sum_{j\in S} j^2 \qquad \text{for all } S \subset N.$$

From these two interpretations we derive that the game v is convex. By (3.1) and (2.8), we get in a straightforward way that the upper vector $b^v = (10,16,18)$ and the core

$$C(v) = \{x \in \mathbb{R}^3_+ \mid x_1 + x_2 + x_3 = 22,\ x_1 \le b^v_1,\ x_2 \le b^v_2,\ x_3 \le b^v_3\}$$
$$= \text{conv}\,\{(10,0,12),\ (10,12,0),\ (6,16,0),$$
$$(0,16,6),\ (0,4,18),\ (4,0,18)\}.$$

The core of the game v is drawn in Figure 5.1. Note that the core is a hexagon inside the imputation set. The six extreme points of the core C(v) are the so-called marginal worth vectors in the game v. The term marginal worth vector is explained by the fact that each coordinate represents a marginal contribution of the corresponding player with respect to a certain coalition in the game. For instance, the marginal worth vector (10,0,12) is obtained from the marginal contributions

$v(123) - v(23)$ of player 1 w.r.t. the coalition $\{2,3\}$,
$v(2) - v(\emptyset)$ of player 2 w.r.t. the coalition $\emptyset$ and
$v(23) - v(2)$ of player 3 w.r.t. the coalition $\{2\}$.

In fact, these three marginal contributions are associated with the ordering (2 3 1) of the three players. It is left to the reader to verify that any extreme point of the core is associated with merely one ordering of the player set.

EXAMPLE 3.2. Let the 3-person game v be given by

$$v(i) = 0 \text{ for all } i \in N, \qquad v(N) = 3,$$
$$v(12) = v(13) = 1 \quad \text{and} \quad v(23) = 0.$$

The game v is equal to the game of Example 2.2 whenever the involved production function $f: \{0,1,2\} \to \mathbb{R}$ is defined by $f(0) = 0$, $f(1) = 1$ and $f(2) = 3$. So, the game is generated by the production economy with two peasants, one landowner and the above production function. From this interpretation and the strict convexity of the production function f, we conclude that the game v is convex. We note that the game v is also generated by any bankruptcy problem with the estate $E = 3$ and the three claims $d_1 \geq 3$, $d_2 = d_3 = 2$. We obtain that the upper vector $b^v = (3,2,2)$ and the core

$$C(v) = \{x \in \mathbb{R}^3_+ \mid x_1 + x_2 + x_3 = 3,\ x_1 \leq b^v_1,\ x_2 \leq b^v_2,\ x_3 \leq b^v_3\}$$
$$= \text{conv}\ \{(3,0,0),\ (1,2,0),\ (0,2,1),\ (0,1,2),\ (1,0,2)\}.$$

The core of the game v is drawn in Figure 5.2. We observe that the core only possesses five instead of six extreme points. This is due to the fact that the marginal worth vector (3,0,0) is associated with both the ordering (2 3 1) and (3 2 1) as illustrated in the following table. In view of this, we say the extreme point (3,0,0) of the core has multiplicity 2. It is left to the reader to verify that the other four extreme points of the core have multiplicity 1.

| | marginal contribution of player i w.r.t. $\theta$ | | |
|---|---|---|---|
| ordering $\theta$ | i=1 $\quad x_1=3$ | i=2 $\quad x_2=0$ | i=3 $\quad x_3=0$ |
| (2 3 1) | $v(123) - v(23)$ | $v(2) - v(\emptyset)$ | $v(23) - v(2)$ |
| (3 2 1) | $v(123) - v(23)$ | $v(23) - v(3)$ | $v(3) - v(\emptyset)$ |

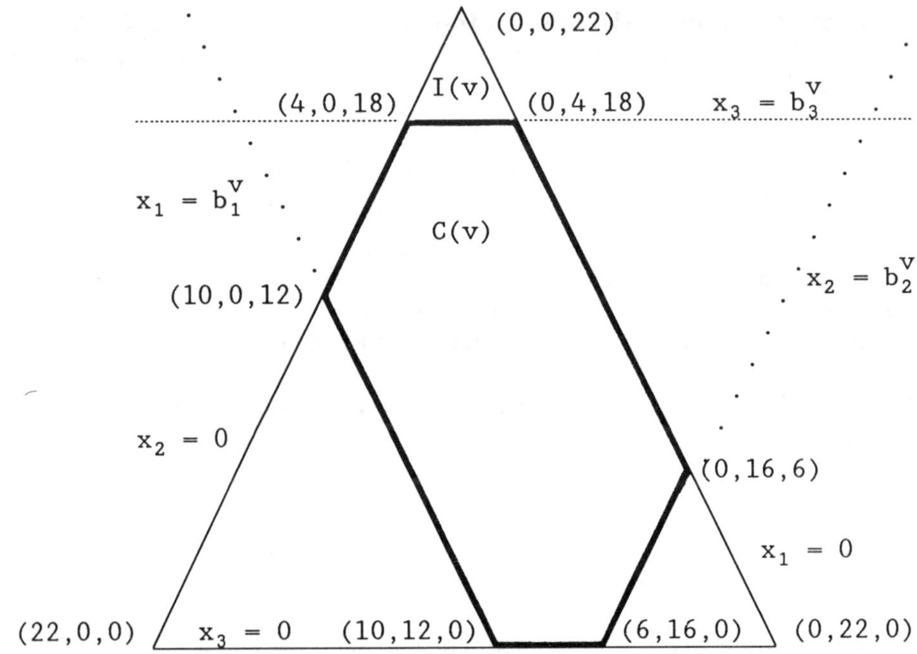


FIGURE 5.1. The indicated hexagon represents the core C(v) of the convex 3-person game v of Example 3.1.

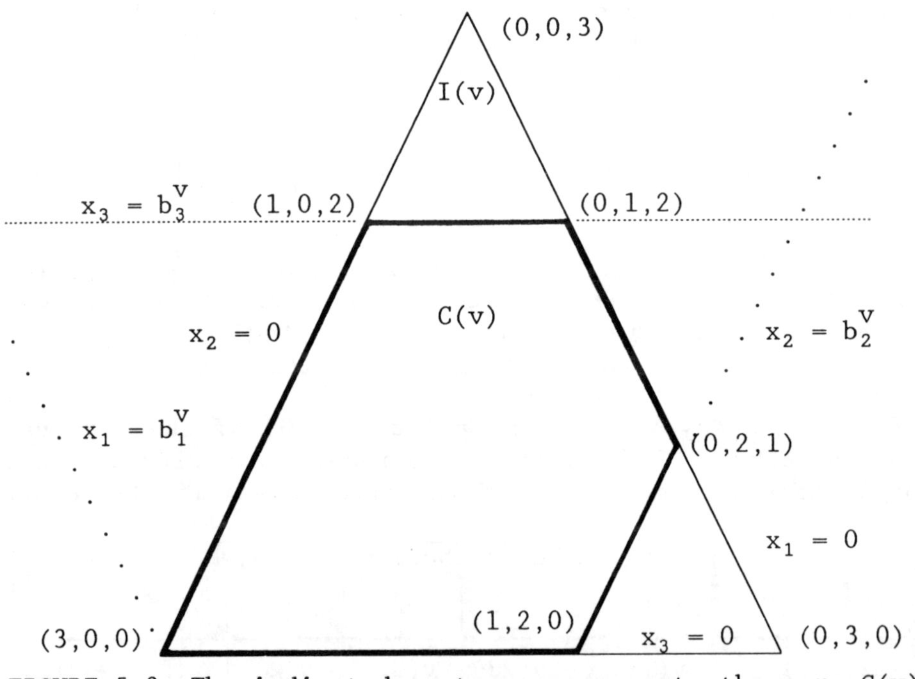


FIGURE 5.2. The indicated pentagon represents the core C(v) of the convex 3-person game v of Example 3.2.

In the above examples we perceived that the core of each involved convex game is the convex hull of the set consisting of the so-called marginal worth vectors whose coordinates represent specific marginal contributions of the corresponding players with respect to the permutations on the player set. Now we present the notion of a marginal worth vector in an accurate way. As usual, the set of all permutations on $N = \{1,2,\ldots,n\}$ is denoted by $\Theta^n$.

DEFINITION 3.3. Let $v \in G^n$ and $\theta \in \Theta^n$.
The <u>marginal</u> <u>worth</u> <u>vector</u> $x^\theta(v) \in \mathbb{R}^n$ with respect to the ordering $\theta$ in the game v is given by

$$x_i^\theta(v) := v(P_i^\theta \cup \{i\}) - v(P_i^\theta) \quad \text{for all } i \in N \qquad (5.10)$$

where $\quad P_i^\theta \quad := \{j \in N \mid \theta(j) < \theta(i)\}.$

Here the set $P_i^\theta$ consists of players who precede player i with respect to the ordering $\theta$ of the player set. Thus, the i-th coordinate $x_i^\theta(v)$ of the marginal worth vector $x^\theta(v)$ represents the marginal contribution of player i to the coalition of his predecessors with respect to the ordering $\theta$ of the player set. According to the parts (i)-(ii) of the next proposition, any marginal worth vector in an n-person game is an efficient payoff vector that satisfies at least n coalition constraints for the core of the game. Further, part (iii) states that a marginal worth vector is either a point outside the core or a vertex of the core. In general, a marginal worth vector is not included in the core of the game as illustrated by the next example.

PROPOSITION 3.4. Let $v \in G^n$ and $\theta \in \Theta^n$.

(i) If $i \in N$ and $S = \{\theta^{-1}(j) \mid 1 \le j \le i\}$, then $\sum_{j \in S} x_j^\theta(v) = v(S)$.

(ii) $x^\theta(v) \in I^*(v)$.

(iii) If $x^\theta(v) \in C(v)$, then $x^\theta(v) \in \text{ext}\, C(v)$.

PROOF. For any player $i \in N$ we define the coalition $S_i \subset N$ by $S_i := \{\theta^{-1}(j) \mid 1 \le j \le i\}$. Put $S_0 := \emptyset$. In view of the formula (5.10), we obtain that for all $i \in N$

$$P^{\theta}_{\theta^{-1}(i)} = \{j \in N \mid \theta(j) < i\} = S_{i-1} \quad \text{and}$$

$$x^{\theta}_{\theta^{-1}(i)}(v) = v(S_{i-1} \cup \{\theta^{-1}(i)\}) - v(S_{i-1})$$

$$= v(S_i) - v(S_{i-1}). \qquad \text{Now it follows that}$$

$$\sum_{j \in S_i} x^{\theta}_j(v) = \sum_{j=1}^{i} x^{\theta}_{\theta^{-1}(j)}(v) = \sum_{j=1}^{i} \left(v(S_j) - v(S_{j-1})\right)$$

$$= v(S_i) - v(S_0) = v(S_i) \qquad \text{for all } i \in N.$$

So, (i) holds. The statement (ii) is a direct consequence of part (i) and (2.1) by noting that $S_n = N$. In order to prove the remaining statement (iii), we define for any coalition $S \subset N$ the corresponding vector $e^S \in \mathbb{R}^n$ by

$$e^S_i := 1 \qquad \text{if } i \in S$$
$$= 0 \qquad \text{if } i \in N-S.$$

Suppose $x^{\theta}(v) \in C(v)$. By part (i), the coalition constraint for the core is an equality whenever the coalition is of the form $S_i$, $i \in N$. From this and the linear independence of the set $\{e^{S_i} \in \mathbb{R}^n \mid i \in N\}$ of the corresponding vectors, we conclude that $x^{\theta}(v) \in \text{ext } C(v)$ whenever $x^{\theta}(v) \in C(v)$. So, (iii) holds. □

EXAMPLE 3.5. Consider again the 1-convex 3-person game v of Example III.5.6. Then the game v is not convex because $v(12) - v(2) > v(123) - v(23)$. The marginal worth vectors in the game v are as follows:

(3,0,15), (9,0,9), (9,9,0), (3,15,0) and (0,9,9).

Here the last marginal worth vector is associated with two orderings of the player set and moreover, it is the only one of these five vectors which belongs to the core C(v) of the game v. The geometric positions of the marginal worth vectors with respect to the core are drawn in Figure 5.3. In the same figure we observe that the core is included in the smallest convex set generated by the five marginal worth vectors.

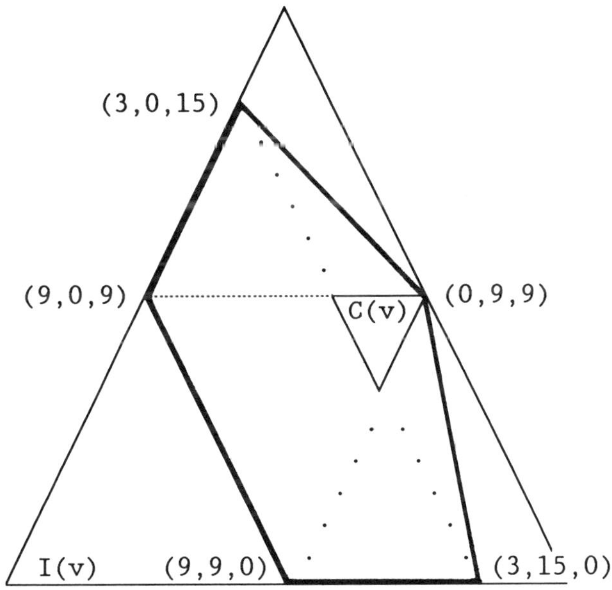


FIGURE 5.3. The geometric positions of the five marginal worth vectors with respect to the core C(v) of the 1-convex 3-person game v of Example 3.5.

The convex hull of the set of all n! marginal worth vectors in an n-person game v is denoted by W(v), i.e.,

$$W(v) := \text{conv}\,\{x^{\theta}(v) \mid \theta \in \Theta^{n}\}. \tag{5.11}$$

Although the marginal worth vectors are in general not closely related to the core of the game v, the set W(v) is a core catcher. The result is due to Weber (1978) whose unpublished proof proceeds by induction on the number n.

THEOREM 3.6. $C(v) \subset \text{conv}\,\{x^{\theta}(v) \mid \theta \in \Theta^{n}\}$ for all $v \in G^{n}$.

As noticed before, a marginal worth vector may fall outside the core, whereas the marginal worth vectors in the games of the Examples 3.1 and 3.2 coincide with the extreme points of the core. For those games, the interrelationship between the marginal worth vectors and the core is due to the convexity of the two games. In fact, for the convexity of an n-person game it is necessary and sufficient that all n! marginal worth vectors belong to the core of the game. In other words, a game v is convex if and only if the core catcher W(v) coincides with the core C(v) of the game v, i.e., the inclusion of Theorem 3.6 is an equality. A third characterization states that a game is convex if and only if the vertices of the core of the game are precisely the marginal worth vectors. These three characterizations of convex games in terms of the marginal worth vectors and the core are listed in the next fundamental theorem concerning convexity.

THEOREM 3.7. The following four statements are equivalent.

(i) $v \in G^n$ is a convex game.

(ii) $x^\theta(v) \in C(v)$ for all $\theta \in \Theta^n$.

(iii) $C(v) = \text{conv}\,\{x^\theta(v) \mid \theta \in \Theta^n\}$.

(iv) $\text{ext}\, C(v) = \{x^\theta(v) \mid \theta \in \Theta^n\}$.

Both the implication (i) ⇒ (ii) and (i) ⇒ (iv) of Theorem 3.7 are due to Shapley (1971), while Ichiishi (1981) proved the converse implication (ii) ⇒ (i). We prove Theorem 3.7 by means of the implications (ii) ⇒ (iii) ⇒ (iv) ⇒ (ii) and the equivalence (i) ⇔ (ii). The following proofs of the two implications (i) ⇒ (ii) and (ii) ⇒ (i) are taken from Ichiishi (1981) because of their uniformity.

PROOF of the equivalence (i) ⇔ (ii) of Theorem 3.7.
(a) Suppose that (i) holds. Let $\theta \in \Theta^n$ be an ordering of N. Then we have $x^\theta(v) \in I^*(v)$ by Proposition 3.4(ii). We prove $x^\theta(v) \in C(v)$ by verifying the coalition constraints for the core of v. Let $S \subset N$, $S \neq \emptyset$. Put $s := |S| \geq 1$ and write $S = \{i_1, i_2, \ldots, i_s\}$ such that $\theta(i_1) < \theta(i_2) < \cdots < \theta(i_s)$. Put $[i_0] := \emptyset$ and $[i_j] := \{i_1, i_2, \ldots, i_j\}$ for all $1 \leq j \leq s$. Now we obtain that $[i_{j-1}] \subset P^\theta_{i_j}$ for all $1 \leq j \leq s$. From the convexity condition (5.2) we deduce that for all $1 \leq j \leq s$

$$v(P^\theta_{i_j} \cup \{i_j\}) - v(P^\theta_{i_j}) \geq v([i_j]) - v([i_{j-1}]).$$

Hence,

$$\sum_{j \in S} x^\theta_j(v) = \sum_{j=1}^{s} x^\theta_{i_j}(v) = \sum_{j=1}^{s} \left(v(P^\theta_{i_j} \cup \{i_j\}) - v(P^\theta_{i_j})\right)$$

$$\geq \sum_{j=1}^{s} \left(v([i_j]) - v([i_{j-1}])\right) = v([i_s]) - v([i_0]) = v(S).$$

Thus, $x^\theta(v) \in C(v)$. So, (i) implies (ii).

(b) To prove the converse implication, suppose that (ii) holds. We verify the convexity condition (5.2). Let $i \in N$, $S \subset N$ and $T \subset N$ be such that $S \subset T \subset N-\{i\}$. Put $s := |S|$ and $t := |T|$. Then $0 \le s \le t \le n-1$ and write

$$S = \{i_1, i_2, \ldots, i_s\}, \qquad T-S = \{i_{s+1}, i_{s+2}, \ldots, i_t\},$$
$$N-T = \{i_{t+1}, i_{t+2}, \ldots, i_n\} \text{ where } i_{t+1} = i.$$

Put $[i_0] := \emptyset$ and $[i_j] := \{i_1, i_2, \ldots, i_j\}$ for all $1 \le j \le n$. Define the permutation $\theta$ on N by $\theta(i_j) := j$ for all $j \in N$. In view of (5.10), we obtain that for all $1 \le j \le n$

$$P^\theta_{i_j} = [i_{j-1}] \quad \text{and} \quad x^\theta_{i_j}(v) = v([i_j]) - v([i_{j-1}]).$$

From this we deduce $x^\theta_{i_{t+1}}(v) = v(T \cup \{i_{t+1}\}) - v(T)$ as well as $\sum_{j\in S} x^\theta_j(v) = \sum_{j=1}^{s} x^\theta_{i_j}(v) = v([i_s]) - v([i_0]) = v(S)$.

Also, $\sum_{j\in S\cup\{i\}} x^\theta_j(v) \ge v(S \cup \{i\})$ since $x^\theta(v) \in C(v)$. Hence,

$$v(S \cup \{i\}) - v(S) \le x^\theta_i(v) = v(T \cup \{i\}) - v(T).$$

Thus, (5.2) holds. So, (ii) implies (i). □

PROOF of the implications (iv) ⇒ (ii) ⇒ (iii) ⇒ (iv) of Theorem 3.7. The implication (iv) ⇒ (ii) trivially holds.
(a) Suppose that (ii) holds. Because the core is a convex set, it follows that $\text{conv}\,\{x^\theta(v) \mid \theta \in \Theta^n\} \subset C(v)$. In view of (5.11) and Theorem 3.6, we obtain $W(v) \subset C(v) \subset W(v)$ and hence, $C(v) = W(v)$. So, (ii) implies (iii).
(b) Suppose that (iii) holds. Define $Y := \{x^\theta(v) \mid \theta \in \Theta^n\}$. Then $Y \subset \text{conv}\,Y = C(v)$ because of (iii). From $Y \subset C(v)$ and Proposition 3.4(iii), we derive the inclusion $Y \subset \text{ext}\,C(v)$. It remains to prove the inverse inclusion $\text{ext}\,C(v) \subset Y$. Let $x \in \text{ext}\,C(v)$. Then we have $x \in C(v) = \text{conv}\,Y$ and therefore,

$$x = \sum_{\theta\in\Theta^n} \alpha_\theta x^\theta(v) \text{ where } \alpha_\theta \ge 0 \text{ for all } \theta \in \Theta^n \text{ and } \sum_{\theta\in\Theta^n} \alpha_\theta = 1.$$

Moreover, $x^\theta(v) \in C(v)$ for all $\theta \in \Theta^n$ because $Y \subset C(v)$. Now $x \in \text{ext}\,C(v)$ yields $x = x^\theta(v)$ for all $\theta \in \Theta^n$ with $\alpha_\theta \ne 0$. Hence, $x \in Y$ whenever $x \in \text{ext}\,C(v)$. Thus, $\text{ext}\,C(v) \subset Y$ and consequently, $Y = \text{ext}\,C(v)$. So, (iii) implies (iv). □

Figure 5.3 illustrates that the core of a nonconvex game may possess vertices which are interior to the imputation set of the game. For any convex n-person game v, no vertex of the core is interior to the imputation set I(v) since all n! marginal worth vectors are boundary points of the imputation set because of the equality

$$x_i^\theta(v) = v(\{i\}) \quad \text{whenever } \theta \in \Theta^n,\ i \in N \text{ such that } \theta(i) = 1.$$

The geometric positions of the vertices of the core of a convex 3-person game with respect to the imputation set are drawn in Figure 5.4.

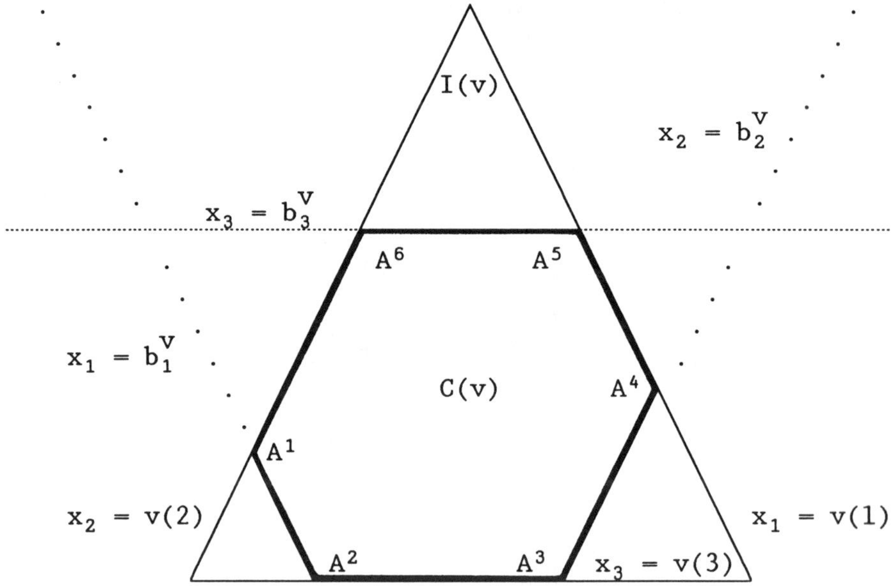


$$A^1 = \left(b_1^v,\ v(2),\ v(23) - v(2)\right) \text{ w.r.t. } \theta = (2\ 3\ 1)$$

$$A^2 = \left(b_1^v,\ v(23) - v(3),\ v(3)\right) \text{ w.r.t. } \theta = (3\ 2\ 1)$$

$$A^3 = \left(v(13) - v(3),\ b_2^v,\ v(3)\right) \text{ w.r.t. } \theta = (3\ 1\ 2)$$

$$A^4 = \left(v(1),\ b_2^v,\ v(13) - v(1)\right) \text{ w.r.t. } \theta = (1\ 3\ 2)$$

$$A^5 = \left(v(1),\ v(12) - v(1),\ b_3^v\right) \text{ w.r.t. } \theta = (1\ 2\ 3)$$

$$A^6 = \left(v(12) - v(2),\ v(2),\ b_3^v\right) \text{ w.r.t. } \theta = (2\ 1\ 3)$$

FIGURE 5.4. The geometric positions of the marginal worth vectors $A^i$, $1 \le i \le 6$, in a convex three-person game v with respect to the imputation set I(v).

Obviously, the core of a convex game degenerates into a singleton whenever all marginal worth vectors coincide. As an example, we consider the convex unanimity n-person game $u_T$ of (1.5) with only one veto player i, i.e., $T = \{i\}$ and

$$u_i(S) = 1 \quad \text{if } i \in S$$
$$= 0 \quad \text{if } i \in N-S.$$

In a straightforward way we derive from the formula (5.10) that all marginal worth vectors in the unanimity n-person game $u_i$ coincide with the i-th unit vector $e^i \in \mathbb{R}^n$ and so, $C(u_i) = \{e^i\}$ by Theorem 3.7. Notice that also $I(u_i) = \{e^i\}$. For simple games, we have that any marginal worth vector is an efficient payoff vector whose coordinates are equal to either zero or one. From this fact, it follows immediately that the set of all n! marginal worth vectors in a simple n-person game is included in the set of all n unit vectors of $\mathbb{R}^n$. In particular, it is left to the reader to verify that the marginal worth vectors in the convex unanimity n-person games $u_T$, $T \subset N$, $T \neq \emptyset$, of (1.5) are as follows:

$$x^\theta(u_T) = e^i \quad \text{iff} \quad i \in T \text{ and } \theta(i) = \max\big(\theta(j) \mid j \in T\big).$$

Therefore, $C(u_T) = \text{conv}\,\{e^i \mid i \in T\}$ for all $T \subset N$, $T \neq \emptyset$. Remark that the multiplicity of each vertex $e^i$, $i \in T$, of the core of the unanimity n-person game $u_T$ equals $|T|^{-1}n!$.

## 4. The Shapley value and the $\tau$-value of a convex game

The Shapley value of any game can be calculated in a straightforward way with the aid of the well-known formula (2.3). For the three games of the Examples 3.1, 3.2 and 3.5 respectively, we deduce from the formula (2.3) that their Shapley values are equal to (5,8,9), $\frac{1}{6}(8,5,5)$ and (4,7,7). Furthermore, we observe that the Shapley value of the game of Example 3.5 lies outside the core, whereas the Shapley value payoff vector of each of the other two games even coincides with the centre of gravity of the "six" extreme points of the core where a certain multiplicity is taken into account. The central position of the Shapley value within the core is due to the convexity of the games of the Examples 3.1 and 3.2. In order to prove the general result concerning the Shapley value of a convex game, we introduce a second alternative formula for the Shapley value which was already mentioned in Shapley (1953).

PROPOSITION 4.1.
The Shapley value $\phi(v) \in \mathbb{R}^n$ of a game $v \in G^n$ is as follows: for all $i \in N$

$$\phi_i(v) = (n!)^{-1} \sum_{\theta \in \Theta^n} \left(v(P_i^\theta \cup \{i\}) - v(P_i^\theta)\right). \tag{5.12}$$

Shortly,

$$\phi(v) = (n!)^{-1} \sum_{\theta \in \Theta^n} x^\theta(v) \qquad \text{for all } v \in G^n. \tag{5.13}$$

PROOF. We prove that the formula (5.12) equals the familiar formula (2.3). Let $v \in G^n$ and $i \in N$. Obviously, the terms

$v(P_i^\theta \cup \{i\}) - v(P_i^\theta),\ \theta \in \Theta^n,$ are of the form

$v(S \cup \{i\}) - v(S),\quad S \subset N-\{i\}.$

For any $T \subset N-\{i\}$ we determine the number of permutations $\theta$ on N such that the equality $P_i^\theta = T$ holds. For $P_i^\theta = T$, it is necessary and sufficient that the permutation $\theta$ satisfies

$$\theta(i) = |T| + 1,$$
$$\theta(j) \le |T| \qquad \text{for all } j \in T,$$
$$\theta(j) \ge |T| + 2 \qquad \text{for all } j \notin T \cup \{i\}.$$

Now it follows that there are precisely $|T|!(n-|T|-1)!$ permutations $\theta$ on N with $P_i^\theta = T$. Hence, the formula (5.12) is equal to the familiar formula (2.3). □

THEOREM 4.2. The Shapley value $\phi(v)$ of a convex game v is the centre of gravity of the extreme points of the core C(v) of the game v where the multiplicity of the extreme points of the core is taken into account. In particular,

$$\phi(v) \in C(v) \qquad \text{for all } v \in C^n.$$

The above theorem was already noted in Shapley (1971) as an adjunct to the study of the core of a convex game. Theorem 4.2 follows directly from Theorem 3.7 and formula (5.13). From Theorem 4.2 we deduce that a game is not convex when the Shapley value of the game does not belong to the core.

Finally, we call attention to the $\tau$-value of a convex game v which represents some efficient compromise between the upper bound $b^v$ and the lower bound $b^v - \lambda^v$ for the core of the game. By Proposition 1.3(i), the gap function $g^v$ of a convex game v is monotonic and therefore, the associated concession vector $\lambda^v$ is determined by the gaps of the one-person coalitions. Thus, the convexity of a game v implies

$$b_i^v - \lambda_i^v = b_i^v - g^v(\{i\}) = v(\{i\}) \qquad \text{for all } i \in N.$$

The upper vector $b^v$ and the disagreement vector $b^v - \lambda^v$ of a convex n-person game v are bounds for the core which are as sharp as possible because the position of the marginal worth vectors $x^\theta(v)$, $\theta \in \Theta^n$, within the core is such that

$$x_i^\theta(v) = v(\{i\}) - v(\emptyset) = v(\{i\}) \qquad \text{whenever } \theta(i) = 1,$$

$$x_i^\theta(v) = v(N) - v(N-\{i\}) = b_i^v \qquad \text{whenever } \theta(i) = n.$$

We conclude that the two vectors $b^v$ and $b^v - \lambda^v$ are sharp bounds for the core if the game v is convex. Nevertheless, the $\tau$-value of a convex game may fall outside the core.

EXAMPLE 4.3. Consider the numerical airport cost game (N;c) with five landings by planes of three different types such that $|N_1| = 1$, $|N_2| = |N_3| = 2$ and $C_1 = 8$, $C_2 = 10$, $C_3 = 36$. Due to Theorem IV.4.2(i), the $\tau$-value cost allocation $\tau(c)$ for this airport cost game is given by

$$\tau_i(c) = 0.36C_j \qquad \text{whenever } i \in N_j. \text{ So,}$$

$$\sum_{j \in S} \tau_j(c) = 10.08 > 10 = c(S) \qquad \text{for } S = N_1 \cup N_2.$$

Let (N;v) be the associated savings game. Then it follows from Example 2.4 that v is a convex savings game, while the above strict inequality is equivalent to

$$\sum_{j \in S} \tau_j(v) < v(S) \qquad \text{for } S = N_1 \cup N_2.$$

Consequently, $\tau(v) \notin C(v)$ although the game v is convex.

As already mentioned at the end of Section 1, the $\tau$-value of a convex game can usually be determined with the aid of Theorem III.6.2 since convexity implies semiconvexity. In particular, the $\tau$-value of a zero-normalized convex game is proportional to the upper vector of the game. As a result, the $\tau$-values of the two convex games of the Examples 3.1 and 3.2 respectively are equal to (5,8,9) and $\frac{3}{7}(3,2,2)$.

## 5. The stable set of a convex game

Stable sets are defined by means of an internal and external stability condition. In view of Theorem II.4.2(i), the core is always internally stable because core-elements can not be dominated by any imputation. Here the external stability condition for the core $C(v)$ of a game $v \in G^n$ is as follows:

for every $y \in I(v) - C(v)$ there exist

$x \in C(v)$ and $T \subset N$, $T \neq \emptyset$, such that

$x$ dominates $y$ by using $T$, i.e.,

$$x_i > y_i \text{ for all } i \in T \quad \text{and} \quad x(T) = v(T). \tag{5.14}$$

We first investigate the external stability of the core of convex three-person games by paying special attention to the core structure as drawn in the two Figures 5.4 and 5.5.

EXAMPLE 5.1. We consider a convex three-person game $v$. Then

$$C(v) = \{x \in I(v) \mid x_i \le b_i^v \text{ for all } i \in N\},$$

$$I(v) - C(v) = \bigcup_{i \in N} \{y \in I(v) \mid y_i > b_i^v\}.$$

Let $y \in I(v) - C(v)$, say $y \in I(v)$ such that $y_1 > b_1^v$. Define the two points $B^1 \in \mathbb{R}^3$ and $B^2 \in \mathbb{R}^3$ by

$$B^1 = (b_1^v, y_2, v(23) - y_2),$$

$$B^2 = (b_1^v, v(23) - y_3, y_3).$$

From the Figures 5.4 and 5.5 we deduce that the straight line segment with end points $B^1$ and $B^2$ is included in the core of the convex game $v$. For any core-element $x$ on this line segment satisfying $x \neq B^1$ and $x \neq B^2$, we have

$$x_2 > y_2,\ x_3 > y_3,\ x(23) = v(N) - x_1 = v(N) - b_1^v = v(23)$$

and so, $x \in C(v)$ dominates $y$ by using the coalition $\{2,3\}$. Hence, (5.14) holds whenever $y \in I(v)$ such that $y_1 > b_1^v$. The remaining two cases $y_2 > b_2^v$ and $y_3 > b_3^v$ respectively can be treated in a similar way and are left to the reader. We conclude that the core of a convex three-person game is externally stable.

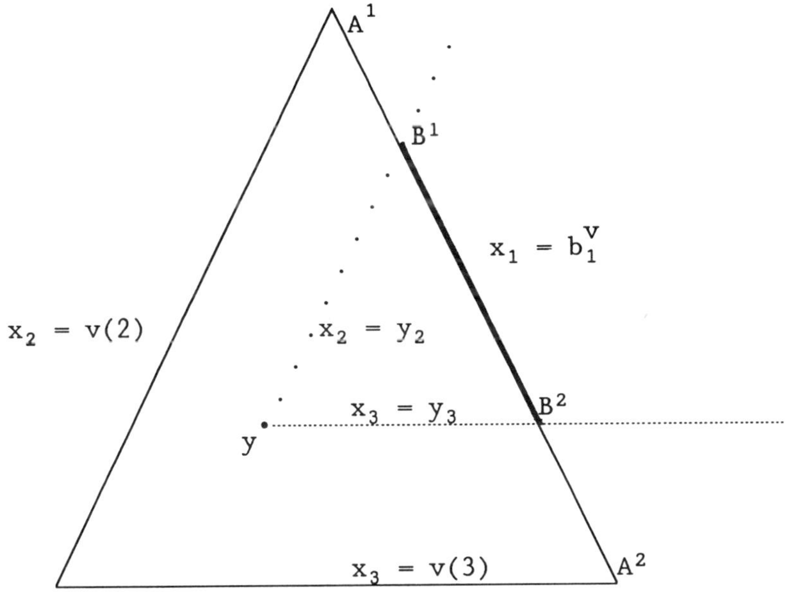


$A^1 = \left(b_1^v, v(2), v(23) - v(2)\right)$ w.r.t. $\theta = (2\ 3\ 1)$,

$A^2 = \left(b_1^v, v(23) - v(3), v(3)\right)$ w.r.t. $\theta = (3\ 2\ 1)$,

$B^1 = (b_1^v, y_2, v(23) - y_2)$, $B^2 = (b_1^v, v(23) - y_3, y_3)$,

$A^1 \in C(v)$, $A^2 \in C(v)$, $y \in I(v)$ such that $y_1 > b_1^v$.

FIGURE 5.5. A detailed form of Figure 5.4.

Shapley (1971) made use of the convexity condition (5.4) in explicit terms to establish that the core of a convex game is a stable set and consequently, the core is the unique stable set for a convex game.

THEOREM 5.2. The core $C(v)$ of a convex game $v$ is the unique stable set for the game.

PROOF. In view of Corollary II.4.3(ii), it is sufficient to prove that the core of a convex game is a stable set. For that purpose, we show that the external stability condition (5.14) for the core of a game holds whenever the game is supposed to be convex. Let both $v \in G^n$ and $y \in I(v) - C(v)$. Define the real number $\alpha$ by

$$\alpha := \max\left(|S|^{-1}[v(S) - y(S)] \,\middle|\, S \subset N,\ S \neq \emptyset\right)$$

and let T be any nonempty coalition for which the maximum $\alpha$ is attained. Then $\alpha > 0$ since $y \in I(v) - C(v)$ and further, $T \neq \emptyset$, $T \neq N$, such that $v(T) - y(T) = |T|\alpha$. Let $\theta: N \to N$ be any permutation such that $T = \{\theta^{-1}(k) \mid 1 \leq k \leq |T|\}$. The corresponding marginal worth vector $x^\theta(v) \in I^*(v)$ satisfies the equality $\sum_{j \in T} x_j^\theta(v) = v(T)$ by using Proposition 3.4(i). Define the vector $x \in \mathbb{R}^n$ by

$$\begin{aligned} x_i &:= y_i + \alpha && \text{if } i \in T \\ &= x_i^\theta(v) && \text{if } i \in N-T. \end{aligned}$$

Thus, $x(T) = y(T) + |T|\alpha = v(T)$ and $x_i > y_i$ for all $i \in T$ because $\alpha > 0$. Furthermore, $x \in I^*(v)$ because

$$\begin{aligned} x(N) &= \sum_{j \in N-T} x_j^\theta(v) + x(T) = v(N) - \sum_{j \in T} x_j^\theta(v) + x(T) \\ &= v(N) - v(T) + x(T) = v(N). \end{aligned}$$

Now we assert that the convexity of v implies $x \in C(v)$. Suppose $v \in C^n$. Then $x^\theta(v) \in C(v)$ by Theorem 3.7(i)-(ii). We establish that $x(S) \geq v(S)$ for all $S \subset N$. Let $S \subset N$. In case $S \cap T = \emptyset$, then we have $x(S) = \sum_{j \in S} x_j^\theta(v) \geq v(S)$ where the involved inequality follows from the fact $x^\theta(v) \in C(v)$. It remains to consider the case $S \cap T \neq \emptyset$. Then we deduce

$$x(S \cap T) = y(S \cap T) + |S \cap T|\alpha \geq v(S \cap T) \qquad \text{and}$$

$$\sum_{j \in S-T} x_j^\theta(v) = \sum_{j \in S \cup T} x_j^\theta(v) - v(T) \geq v(S \cup T) - v(T)$$

where the two inequalities follow from the definition of the real number $\alpha$ and $x^\theta(v) \in C(v)$ respectively. Summing up both inequalities, we obtain

$$x(S) \geq v(S \cap T) + v(S \cup T) - v(T).$$

From the last inequality and the convexity condition (5.4), we derive $x(S) \geq v(S)$. Hence, $x \in C(v)$. We conclude that there exist a coalition $T \subset N$, $T \neq \emptyset$, and a vector $x \in \mathbb{R}^n$ such that $x \in C(v)$, $x_i > y_i$ for all $i \in T$ and $x(T) = v(T)$. So, (5.14) holds. □

## 6. The bargaining set $\mathcal{M}$ of a convex game

We first investigate the interrelationship between the bargaining set $\mathcal{M}$ and the core of convex three-person games. As noticed in Section II.5, the core is always included in the bargaining set $\mathcal{M}$ because there are no objections with respect to any core-element possible.

EXAMPLE 6.1. We consider a convex three-person game v. Then

$$C(v) = \{y \in I(v) \mid y_i \le b_i^v \text{ for all } i \in N\}.$$

We show $I(v) - C(v) \subset I(v) - \mathcal{M}(v)$. Let $x \in I(v) - C(v)$, say $x \in I(v)$ such that $x_1 > b_1^v$. By using (1.6), $x \in I^*(v)$ and (3.1) respectively, we obtain

$$e^v(23,x) = v(23) - x(23) = v(23) - v(N) + x_1 = x_1 - b_1^v > 0.$$

From the positivity of the excess $e^v(23,x)$ we conclude that player 2 can object against player 1 with respect to the imputation x by using the coalition $\{2,3\}$. However, there exists no counterobjection to this objection because

$$e^v(1,x) = v(1) - x_1 < v(1) - b_1^v \le 0,$$
$$e^v(13,x) = v(13) - x(13) < v(13) - v(3) - b_1^v \le 0$$

where the strict inequalities follow from $x_1 > b_1^v$, $x \in I(v)$ and the weak inequalities follow from the known convexity condition (5.2). Therefore, $x \notin \mathcal{M}(v)$ whenever $x \in I(v)$ such that $x_1 > b_1^v$. The remaining two cases $x_2 > b_2^v$ and $x_3 > b_3^v$ respectively can be treated in a similar way and are left to the reader. We have shown that the relevant inclusion $I(v) - C(v) \subset I(v) - \mathcal{M}(v)$ holds and hence, the equality $\mathcal{M}(v) = C(v)$ holds for a convex three-person game v.

Maschler, Peleg and Shapley (1972) established that the bargaining set $\mathcal{M}$ of a convex game coincides with the core of the game. Their proof is essentially based on the notion of the monotonic cover of a game which can be regarded as the least monotonic game that majorizes the original game. Furthermore, the convexity of a game implies the convexity of the monotonic cover according to part (iv) of the next lemma. The subsequent theorem states the result concerning the bargaining set $\mathcal{M}$ of convex games.

LEMMA 6.2. Let the <u>monotonic</u> <u>cover</u> $\acute{v} \in G^n$ of a game $v \in G^n$ be given by

$$\acute{v}(S) := \max\{v(R) \mid R \subset S\} \qquad \text{for all } S \subset N. \tag{5.15}$$

The following statements for a game $v \in G^n$ hold.

(i) $\acute{v}(S) \geq v(S)$ for all $S \subset N$.

(ii) $\acute{v}(S) \leq \acute{v}(T)$ for all $S \subset T \subset N$.

(iii) If another game $w \in G^n$ satisfies $v(S) \leq w(S) \leq w(T)$ for all $S \subset T \subset N$, then $w(S) \geq \acute{v}(S)$ for all $S \subset N$.

(iv) If $v \in C^n$, then $\acute{v} \in C^n$.

PROOF. The statements (i) and (ii) follow immediately from the formula (5.15). In order to prove (iii), suppose that $w \in G^n$ satisfies $v(S_1) \leq w(S_1) \leq w(S_2)$ for all $S_1 \subset S_2 \subset N$. Let $S \subset N$. Then $v(R) \leq w(R) \leq w(S)$ for all $R \subset S$ and hence, $\acute{v}(S) = \max\{v(R) \mid R \subset S\} \leq w(S)$. So, (iii) holds.
It remains to prove the statement (iv). Suppose $v \in C^n$. We verify the convexity condition (5.4) for the game $\acute{v} \in G^n$. Let $S \subset N$ and $T \subset N$. By (5.15), there exist $S_1 \subset S$, $T_1 \subset T$ such that both $\acute{v}(S) = v(S_1)$ and $\acute{v}(T) = v(T_1)$. Now we obtain

$$\acute{v}(S) + \acute{v}(T) = v(S_1) + v(T_1) \leq v(S_1 \cup T_1) + v(S_1 \cap T_1)$$

$$\leq \acute{v}(S_1 \cup T_1) + \acute{v}(S_1 \cap T_1) \leq \acute{v}(S \cup T) + \acute{v}(S \cap T)$$

where the inequalities follow from the convexity condition (5.4) applied to $v$, part (i) and part (ii) respectively. Thus, (5.4) holds for the game $\acute{v}$ and so, $\acute{v} \in C^n$. □

THEOREM 6.3. $\mathcal{M}(v) = C(v)$ for all $v \in C^n$.

PROOF. Let $v \in C^n$. It is sufficient to prove the inclusion $I(v) - C(v) \subset I(v) - \mathcal{M}(v)$. Let $x \in I(v) - C(v)$. Choose $T \subset N$ as a maximal coalition of the largest excess at the imputation $x$, i.e., the coalition $T \subset N$ is chosen such that

$$e^v(T,x) \geq e^v(S,x) \qquad \text{for all } S \subset N,$$

$$e^v(T,x) > e^v(S,x) \qquad \text{for all } S \subset N \text{ with } T \subset S,\ S \neq T.$$

Now $x \in I(v) - C(v)$ implies $e^v(T,x) > 0$ and $2 \leq |T| \leq n-1$. Define the game $w$ with player set $T$ by

$$w(S) := v(S) - x(S) = e^v(S,x) \qquad \text{for all } S \subset T.$$

For its monotonic cover $\acute{w}$ on T we have

$$\acute{w}(T) = \max(w(S) \mid S \subset T) = e^v(T,x) > 0,$$
$$\acute{w}(\{i\}) \geq \acute{w}(\emptyset) = 0 \qquad \text{for all } i \in T.$$

Clearly, the convexity of v implies the convexity of w and therefore, the monotonic cover $\acute{w}$ is also a convex game by Lemma 6.2(iv). It follows from Theorem 3.7 that $C(\acute{w}) \neq \emptyset$. So, there exists $z \in \mathbb{R}^T$ such that $z \in C(\acute{w})$. In particular,

$$z(T) = \acute{w}(T) > 0 \qquad \text{and}$$
$$z_i \geq \acute{w}(\{i\}) \geq 0 \qquad \text{for all } i \in T.$$

From this, it is clear that there exists $\hat{\imath} \in T$ with $z_{\hat{\imath}} > 0$. Define the $|T|$-tuple $\hat{y} = (y_k)_{k \in T}$ of real numbers by

$$y_i := x_i + z_i + \alpha \qquad \text{if } i \in T,\ i \neq \hat{\imath}$$
$$= x_i + z_i - (|T|-1)\alpha \qquad \text{if } i = \hat{\imath}$$

where $\alpha$ is any real number satisfying $0 < \alpha < (|T|-1)^{-1} z_{\hat{\imath}}$. Then

$$y_i > x_i \qquad \text{for all } i \in T \qquad \text{and}$$
$$y(T) = x(T) + z(T) = x(T) + \acute{w}(T)$$
$$= x(T) + e^v(T,x) = v(T).$$

From this we deduce that $(\hat{y};T)$ is an objection of player $\hat{\imath}$ against any player in N−T with respect to the imputation x in the game v.

Now we assert that there exists no counterobjection to the above objection $(\hat{y};T)$. Consider any coalition $R \subset N-\{\hat{\imath}\}$ satisfying $R \cap (N-T) \neq \emptyset$. Because $R \cup T \neq T$, the specific choice of the coalition T yields $e^v(R \cup T,x) < e^v(T,x)$. The strict inequality and the convexity of the game v imply

$$e^v(R,x) \leq e^v(R \cap T,x) + e^v(R \cup T,x) - e^v(T,x)$$
$$< e^v(R \cap T,x). \qquad \text{Furthermore,}$$
$$e^v(R \cap T,x) = w(R \cap T) \leq \acute{w}(R \cap T) \leq z(R \cap T)$$

where the last inequality follows from $z \in C(\acute{w})$. We obtain

$$e^v(R,x) < e^v(R \cap T,x) \leq z(R \cap T)$$
$$= y(R \cap T) - x(R \cap T) - |R \cap T|\alpha$$
$$\leq y(R \cap T) - x(R \cap T) \qquad \text{and hence,}$$
$$v(R) = e^v(R,x) + x(R) < y(R \cap T) + x(R-T).$$

Due to (2.12), the last strict inequality expresses that the coalition R can not be used for a counterobjection. Since R is an arbitrary coalition containing at least one player in N−T but not player $\hat{\imath}$, we conclude that there exists no counterobjection to the above objection $(\hat{y};T)$ of player $\hat{\imath}$ against any player in N−T with respect to the imputation x in the game v. Therefore, $x \notin \mathcal{M}(v)$ as was to be shown. □

## 7. The kernel and the prekernel of a convex game

Due to the fact that the prekernel can be regarded as a simplified version of the kernel, we always have that any imputation in the prekernel is in the kernel as well. But even for the case that the kernel and the prekernel differ, the prekernel gives us some useful information about the kernel. For instance, Theorem II.6.4 stated that the parts of the kernel and the prekernel inside the core do always coincide. In this section we show that zero-monotonicity is a sufficient condition for the kernel to coincide with the prekernel. Here the class of zero-monotonic n-person games is denoted by $M_0^n$, i.e.,

$$M_0^n := \{v \in G^n \mid v(S) + \sum_{j \in T-S} v(\{j\}) \leq v(T) \text{ for all } S \subset T \subset N\}.$$

Obviously, the zero-monotonicity of a game can be described as the intersection of a finite number of weak inequalities that are linear in the worths of the coalitions in the game. Therefore, the class $M_0^n$ is a polyhedral cone in the linear space $G^n$. Since convexity implies zero-monotonicity, we conclude from Theorem 1.2 that the class $M_0^n$ is a full-dimensional cone in $G^n$.

COROLLARY 7.1. $C^n \subset M_0^n$ and $\dim M_0^n = 2^n - 1$.

Throughout the remainder of this section we use the following notation. Let $v \in G^n$ where $n \geq 2$ and $x \in I^*(v)$. By noting (2.28), the set $\Sigma_1(x)$ consists of the nontrivial coalitions of maximum excess at the pre-imputation x among the nontrivial coalitions. Furthermore, we denote by $\mathcal{P}(x)$ the set of all players who are included in any nontrivial coalition of maximum excess at x. Thus,

$$\Sigma_1(x) := \{S \in \Sigma^0 \mid e^v(S,x) \geq e^v(R,x) \text{ for all } R \in \Sigma^0\},$$
$$\mathcal{P}(x) := \bigcap\{S \mid S \in \Sigma_1(x)\} \qquad \text{where}$$
$$\Sigma^0 := \{S \subset N \mid S \neq N, \emptyset\}.$$

Three useful properties of the coalition $\mathcal{P}(x)$ are listed in the next lemma.

LEMMA 7.2. Let $v \in M_0^n$ where $n \geq 2$ and $x \in I^*(v)$.

(i) If $i \in N - \mathcal{P}(x)$, then $x_i \geq v(\{i\})$.

(ii) If there exists $T \in \Sigma_1(x)$ such that $x_i = v(\{i\})$ for all $i \in N-T$, then $\mathcal{P}(x) = \emptyset$.

(iii) If $i \in \mathcal{P}(x)$ and $j \in N - \mathcal{P}(x)$, then $s_{ij}^v(x) > s_{ji}^v(x)$.

PROOF. (i) Suppose $i \in N - \mathcal{P}(x)$. There exists a coalition $T \in \Sigma_1(x)$ such that $i \notin T$. By both (1.6) and $v \in M_0^n$, we get

$$\begin{aligned} &e(T \cup \{i\},x) = v(T \cup \{i\}) - x(T \cup \{i\}) \\ &\geq v(T) - x(T) + v(\{i\}) - x_i \\ &= e(T,x) + e(\{i\},x). \end{aligned} \tag{5.16}$$

If $T \neq N-\{i\}$, then $T \cup \{i\} \neq N,\emptyset$, which implies together with (5.16) as well as $T \in \Sigma_1(x)$ that

$$e(\{i\},x) \leq e(T \cup \{i\},x) - e(T,x) \leq 0.$$

If $T = N-\{i\}$, then we derive from (5.16) that

$$0 = e(N,x) \geq e(T,x) + e(\{i\},x) \geq 2e(\{i\},x)$$

where the last inequality follows from $T \in \Sigma_1(x)$ and $n \geq 2$. In any case we obtain $e(\{i\},x) \leq 0$ which inequality is equivalent to $x_i \geq v(\{i\})$. So, (i) holds.

(ii) Suppose that there exists a coalition $T \in \Sigma_1(x)$ satisfying $x_i = v(\{i\})$ for all $i \in N-T$. Then we deduce

$$\begin{aligned} &e(T,x) = v(T) - x(T) = v(T) + \sum_{j \in N-T} v(\{j\}) - x(N) \\ &\leq v(N) - x(N) = 0 \end{aligned}$$

where the inequality follows from $v \in M_0^n$. As a consequence, $T \in \Sigma_1(x)$ satisfies $e(T,x) \leq 0 = e(\{i\},x)$ for all $i \in N-T$. We conclude that $T \in \Sigma_1(x)$ and $\{i\} \in \Sigma_1(x)$ for all $i \in N-T$. Hence, $\mathcal{P}(x) = \bigcap\{S \mid S \in \Sigma_1(x)\} = \emptyset$. So, (ii) holds.

(iii) Suppose $i \in \mathcal{P}(x)$ and $j \in N - \mathcal{P}(x)$. Then there exists $T \in \Sigma_1(x)$ such that $j \notin T$. Since $i \in \mathcal{P}(x)$ is equivalent to $i \in S$ for all $S \in \Sigma_1(x)$, we obtain that $i \in T$ and $S \notin \Sigma_1(x)$ for all $S \subset N-\{i\}$ as well. From this and the maximum excess of T at x, we conclude that

$$s_{ij}(x) = \max\big(e(S,x) \mid S \in \Gamma_{ij}\big) = e(T,x),$$
$$s_{ji}(x) = \max\big(e(S,x) \mid S \in \Gamma_{ji}\big) < e(T,x). \quad \text{Thus,}$$
$$s_{ij}(x) = e(T,x) > s_{ji}(x). \qquad \text{So, (iii) holds.}$$ □

THEOREM 7.3. $\mathcal{K}(v) = \mathcal{K}^*(v)$ for all $v \in M_0^n$.

PROOF. The equality $\mathcal{K}(v) = I^*(v) = \mathcal{K}^*(v)$ trivially holds for all 1-person games v. Let $v \in M_0^n$ where $n \geq 2$. We prove the inclusions $\mathcal{K}^*(v) \subset \mathcal{K}(v)$ and $\mathcal{K}(v) \subset \mathcal{K}^*(v)$ respectively.
(i) Suppose $x \in \mathcal{K}^*(v)$. By (2.15), we have $x \in I^*(v)$ such that $s_{ij}(x) = s_{ji}(x)$ for all $i,j \in N$, $i \neq j$. Together with Lemma 7.2(iii) and $\mathcal{P}(x) \neq N$, this implies $\mathcal{P}(x) = \emptyset$. It follows from Lemma 7.2(i) that $x_i \geq v(\{i\})$ for all $i \in N$ and thus, $x \in I(v)$. Consequently, $x \in \mathcal{K}^*(v) \cap I(v) \subset \mathcal{K}(v)$. We have shown that the first inclusion $\mathcal{K}^*(v) \subset \mathcal{K}(v)$ holds.
(ii) In order to prove the inverse inclusion $\mathcal{K}(v) \subset \mathcal{K}^*(v)$, suppose $x \in \mathcal{K}(v)$. First of all, we show $\mathcal{P}(x) = \emptyset$.
Assume $\mathcal{P}(x) \neq \emptyset$. Choose $i \in \mathcal{P}(x)$ as well as $T \in \Sigma_1(x)$. We deduce from Lemma 7.2(ii) that there exists $j \in N-T$ with $x_j > v(\{j\})$. In particular, we have $j \in N - \mathcal{P}(x)$ and hence, $s_{ij}(x) > s_{ji}(x)$ by using Lemma 7.2(iii). It follows that $\big(s_{ij}(x) - s_{ji}(x)\big)\big(x_j - v(\{j\})\big) > 0$ which strict inequality is in contradiction with $x \in \mathcal{K}(v)$. We have shown $\mathcal{P}(x) = \emptyset$.
Now we prove $x \in \mathcal{K}^*(v)$. Assume $x \notin \mathcal{K}^*(v)$. By (2.15), there exist $i,j \in N$, $i \neq j$, such that $s_{ij}(x) > s_{ji}(x)$. From this and $x \in \mathcal{K}(v)$ as well, it follows that $x_j = v(\{j\})$. Further, $\mathcal{P}(x) = \emptyset$ implies that there exists $T \in \Sigma_1(x)$ with $i \notin T$. We get $e(T,x) \geq \max\big(e(S,x) \mid S \in \Gamma_{ij}\big) > \max\big(e(S,x) \mid S \in \Gamma_{ji}\big)$ which strict inequality yields $T \notin \Gamma_{ji}$. From $T \notin \Gamma_{ji}$ and $i \notin T$, we derive $j \notin T$. We obtain

$$s_{ij}(x) > s_{ji}(x) \geq e(T \cup \{j\},x)$$
$$= v(T \cup \{j\}) - x(T \cup \{j\}) \geq v(T) - x(T) + v(\{j\}) - x_j$$
$$= v(T) - x(T) = e(T,x)$$

where the last inequality follows from $v \in M_0^n$. However, the strict inequality $e(T,x) < s_{ij}(x)$ is in contradiction with $T \in \Sigma_1(x)$. We conclude that $x \in \mathcal{K}^*(v)$. This completes the proof of the second and final inclusion $\mathcal{K}(v) \subset \mathcal{K}^*(v)$. □

From the above theorem which is due to Maschler, Peleg and Shapley (1972), we deduce that the kernel and the prekernel of a convex game coincide. According to Theorem II.6.3, the kernel is always included in the bargaining set $\mathcal{M}$, while the bargaining set $\mathcal{M}$ of a convex game coincides with the core by Theorem 6.3. As a direct consequence, the kernel of a convex game is a subset of the core. As a matter of fact, Maschler, Peleg and Shapley (1972) proved that the kernel of a convex game is a singleton and as such, the kernel of a convex game coincides with the nucleolus of the game.

THEOREM 7.4. $\mathcal{K}(v) = \mathcal{K}^*(v) = \{\eta(v)\}$ for all $v \in C^n$.

Although the nucleolus of a convex game occupies a central position within the regular core, an explicit formula for the nucleolus of a convex game is not available. The proof of Theorem 7.4 is fairly elaborate and is omitted here. We remark that the proof proceeds by induction on the number of players in the game and is based on the notions of the set $\Sigma_1(x)$ of coalitions and so-called stage games induced by a pre-imputation x for the original game.

Finally, we illustrate the geometric position of the unique kernel-element with respect to the core for the two convex games of the Examples 3.1 and 3.2. For the game v of Example 3.1, the nucleolus equals the centre of gravity of the extreme points of the core, i.e., $\eta(v) = (5,8,9)$ and in addition, the nucleolus coincides with the Shapley value as well as the $\tau$-value payoff vector. A geometric verification of the nucleolus payoff vector is presented in Figure 5.6 which illustrates the bisection property for the payoff vector $x := (5,8,9)$ with respect to the core of the game v. By (2.16), we obtain the following results concerning the largest straight line segments through the point x in the various directions that are still included in the core.

$$\delta^v_{12}(0,x) = 5, \qquad R^v_{12}(0,x) = \text{conv}\,\{(10,3,9),\ (0,13,9)\},$$

$$\delta^v_{13}(0,x) = 5, \qquad R^v_{13}(0,x) = \text{conv}\,\{(10,8,4),\ (0,8,14)\},$$

$$\delta^v_{23}(0,x) = 8, \qquad R^v_{23}(0,x) = \text{conv}\,\{(5,16,1),\ (5,0,17)\}.$$

Clearly, the given core-element x is the midpoint of the three above line segments and hence, $x = \eta(v)$ by the two Theorems II.6.7 and 7.4. The reader is invited to study Figure 5.6.

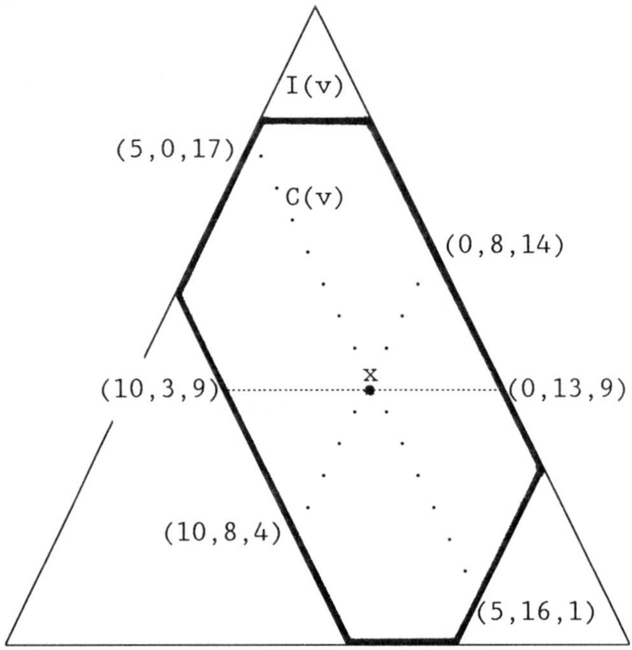


$x := (5,8,9)$

$\delta^v_{12}(0,x) = \delta^v_{21}(0,x) = 5$

$\delta^v_{13}(0,x) = \delta^v_{31}(0,x) = 5$

$\delta^v_{23}(0,x) = \delta^v_{32}(0,x) = 8$

$\eta(v) = (5,8,9)$

FIGURE 5.6. The bisection property for the nucleolus payoff vector $\eta(v)$ with respect to the core C(v) of the convex three-person game v of Example 3.1.

For the game v of Example 3.2, it is left to the reader to verify that the strong $\epsilon$-core $C_\epsilon(v) \neq \emptyset$ iff $\epsilon \geq -1$ and in particular, we get $C_{-1}(v) = \{(1,1,1)\}$. Therefore, we obtain

$$\eta(v) = (1,1,1), \quad \text{while}$$

$$\phi(v) = \frac{1}{6}(8,5,5) \quad \text{and}$$

$$\tau(v) = \frac{3}{7}(3,2,2).$$

The bisection property for the nucleolus payoff vector with respect to the core of the convex game is illustrated in Figure 5.7. The reader is invited to study this figure.

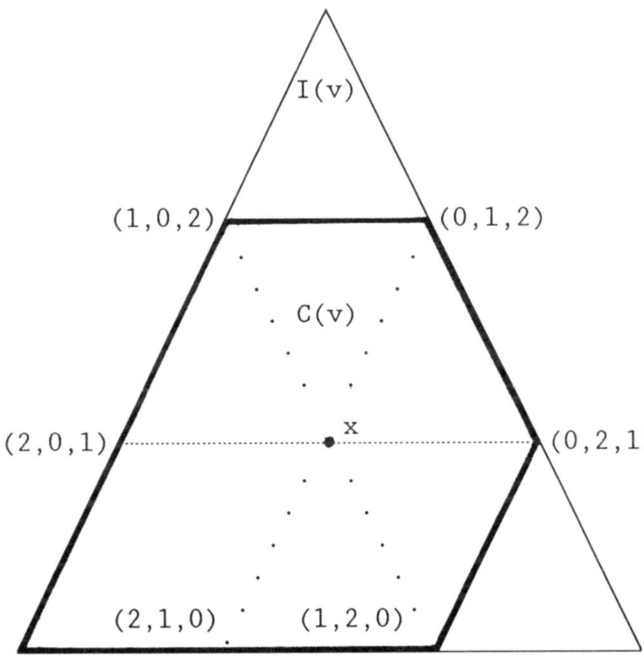


$x := (1,1,1)$

$\delta^v_{ij}(0,x) = 1 \quad$ for all $i,j \in N$, $i \neq j$

$\eta(v) = (1,1,1)$

FIGURE 5.7. The bisection property for the nucleolus payoff vector $\eta(v)$ with respect to the core C(v) of the convex three-person game v of Example 3.2.

In the next chapter, the focus is on a particular type of a convex game which is also known as a bankruptcy game. In Chapter VII we present a generalization of the convexity notion by imposing the well-known convexity condition on a related game instead of the original game itself. Here the related game is obtained by means of an easy transformation on the original game. It may happen that the related game coincides with the original game and hence, the convexity theory can be applied to both games. In case the related game differs from the original game, then the fundamental theorems concerning convexity can be used to obtain similar results for various solution concepts of the original game.

CHAPTER VI

# DIVISION RULES AND ASSOCIATED GAME THEORETIC SOLUTIONS FOR BANKRUPTCY PROBLEMS

## 1. Introduction

Throughout the chapter we pay attention to the general bankruptcy problem as presented in Section I.6. In each of the next three sections we treat one specific division rule for the bankruptcy problem as well as the associated game theoretic one-point solution concept for the corresponding bankruptcy game.
As usual, the estate E and the claims (debts) $d_i$, $i \in N$, are nonnegative real numbers satisfying $\sum_{j\in N} d_j \geq E$. Let $N := \{1,2,\ldots,n\}$ where $n \geq 2$ and $d := (d_1,d_2,\ldots,d_n) \in \mathbb{R}^n$.
The general bankruptcy problem $(E;d)$ generates the n-person game $(N;v_{E;d})$ whose characteristic function is given by

$$v_{E;d}(S) := \max\big(0,\ E - d(N-S)\big) \qquad \text{for all } S \subset N. \quad (6.1)$$

Clearly, we have $v_{E;d}(N) = E$. In Example V.2.3, it is shown that the bankruptcy game of (6.1) is convex.
A <u>division</u> <u>rule</u> is a nonnegative function f that assigns to any general bankruptcy problem $(E;d)$ an efficient payoff vector $f(E;d) \in \mathbb{R}^n_+$, i.e.,

$$\sum_{j\in N} f_j(E;d) = E \qquad \text{and} \qquad f_i(E;d) \geq 0 \text{ for all } i \in N.$$

These rules prescribe somehow a specific distribution of the total estate E among the n creditors in such a way that the payoff to any creditor is at least zero.
The organization of the next four sections is as follows. Section 2 mainly deals with the recursive completion (RC-) division rule which agrees with the game theoretic notion of the Shapley value (O'Neill, 1982). In Section 3 we treat the relationship of the contested garment consistent (CGC-)

division rule to the nucleolus (Aumann and Maschler, 1985). Further, Section 4 compares the adjusted proportional (AP-) division rule with the $\tau$-value (Curiel, Maschler and Tijs, 1987). Finally, Section 5 is devoted to an axiomatization of the CGC- and the AP-division rule.

## 2. The recursive completion division rule in comparison with the Shapley value

The study of the bankruptcy problem was started in O'Neill (1982) who interpreted the bankruptcy problem $(E;d)$ as the problem of n heirs and n corresponding wills such that each will $i \in N$ only mentions the claim $d_i$ for heir i which claims total at least as much as the estate E of the died person. In addition, O'Neill required that each claim is at most the total estate, i.e., $d_i \le E$ for all $i \in N$. The bankruptcy problems satisfying the additional condition are called <u>simple bankruptcy problems</u>.

O'Neill's approach to the simple bankruptcy problem is based on the idea of completing any will by a specific determination of the amounts implicitly left to the other heirs j, $j \in N-\{i\}$, not mentioned in the will $i \in N$. For that purpose, O'Neill introduced the associated n×n claims matrix D whose diagonal entries are equal to the claims $d_1, d_2, \cdots, d_n$ respectively, whereas the other entries are not yet filled in. The rows (columns respectively) of the matrix D are indexed by the wills (heirs). The completion of the claims matrix is performed recursively as follows. In the first step we deduce from the original n×n claims matrix D the $(n-1)\times(n-1)$ matrices $D_i^1$, $i \in N$, by deleting will i and heir i. The diagonal entries of the matrix $D_i^1$ are the truncated claims $\min\left(d_j, E - d_i\right)$, $j \in N-\{i\}$, in the generated simple bankruptcy problem with the new total estate $E - d_i$. In turn, each $(n-1)\times(n-1)$ matrix $D_i^1$ generates a list of $(n-2)\times(n-2)$ matrices $D_{ij}^2$, $j \in N-\{i\}$, by deleting both will j and heir j. The procedure continues recursively until we arrive at a total list of n! simple bankruptcy problems with only one heir. Concerning these one-person problems, the claim of the single heir $k \in N$ and the total estate are equal to $\max\left(0, E - d(N-\{k\})\right)$.

In the second step we solve each one-person bankruptcy problem in a trivial way by giving away the total estate to the single heir. The two blanks of any 2×2 matrix are filled in with the aid of the known solutions for the two generated one-person bankruptcy problems. The solution for each 2×2 matrix is obtained by averaging each heir's claim over the two corresponding wills. The procedure continues recursively in such a way that for any $1 \le k \le n-1$ we solve each completed k×k matrix by averaging each heir's claim over the k corresponding wills and then the blanks of any (k+1)×(k+1) matrix are filled in with the aid of the known solutions for the k×k matrices that are generated by the involved (k+1)×(k+1) matrix as indicated in the first step. Finally, the blanks of the original n×n claims matrix D are completed.

The <u>recursive</u> <u>completion</u> (RC-) <u>rule</u> assigns to any simple n-person bankruptcy problem the payoff vector that is obtained by averaging each heir's claim over the n wills in the associated and completed n×n claims matrix. In point of fact, the resulting payoff vector is efficient since the trivial solution for each one-person problem is efficient, while for any $1 \le k \le n-1$, the efficiency of solutions for k-person problems implies the efficiency of the solution for each (k+1)-person problem.

EXAMPLE 2.1. In order to illustrate the RC-division rule, we consider the simple bankruptcy problem with the estate $E = 400$ and the three claims $d_1 = 100$, $d_2 = 200$, $d_3 = 300$. Figure 6.1 shows the calculations which are necessary for the determination of the division of the estate according to the RC-rule. We obtain $RC(E;d) = \frac{1}{3}(200,350,650)$.

Both the deletion of the heirs in a certain order and the determination of solutions for subproblems by averaging give rise to look closely at all possible orderings of the heirs on the understanding that all orderings are equally likely to occur. Each ordering of N can be regarded as an order of presenting the claims of the heirs at court. The heir who arrives at court first with respect to the given ordering is paid his entire claim, while the next heir receives the remaining part of the estate or his claim, whichever is less, and so on. This procedure yields the same division as the RC-rule since the payoff to the first heir according to his claim reduces the n-person problem to

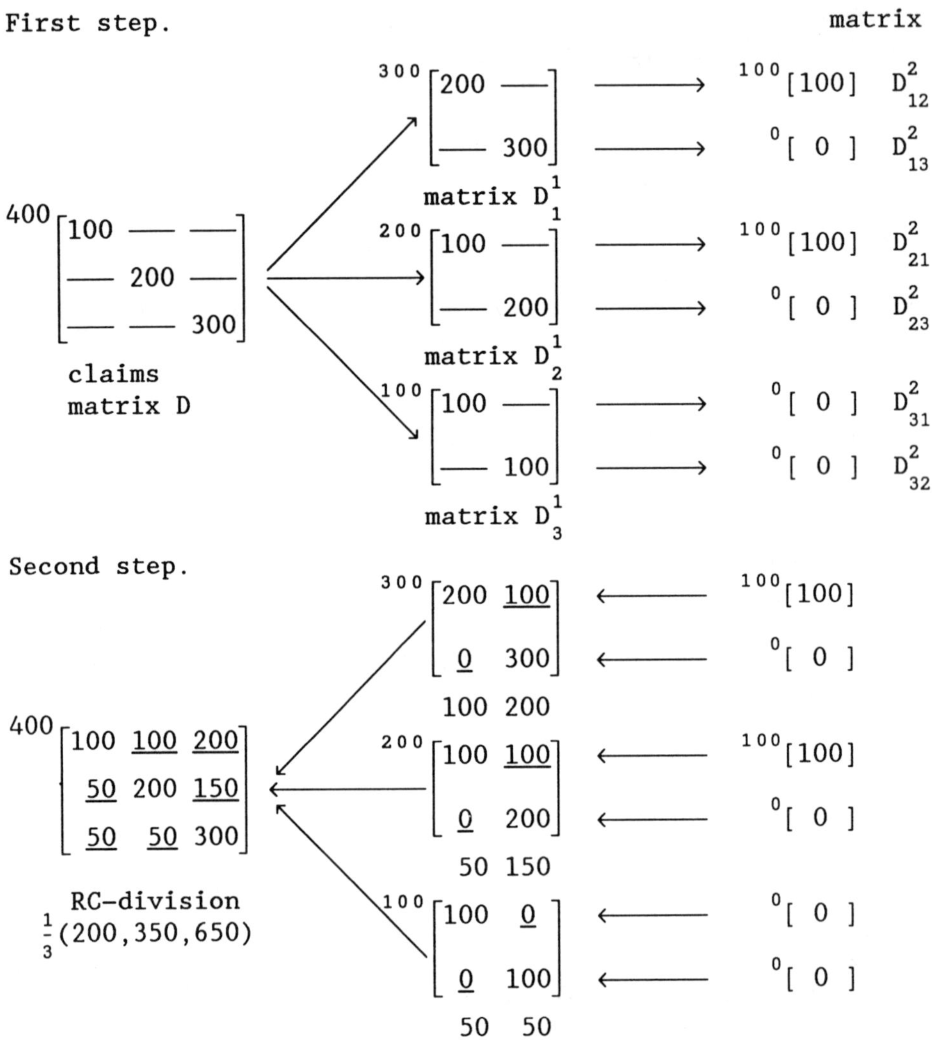


FIGURE 6.1. Necessary calculations for the determination of the recursive completion (RC–) division applied to the simple bankruptcy problem with $E = 400$, $d_1 = 100$, $d_2 = 200$, $d_3 = 300$. The flow of the calculations is indicated by arrows. The first step shows the reduction to one-person problems where the total estates appear to the upper left of each matrix. The second step shows the determination of solutions for subproblems by averaging. The relevant solutions appear below each matrix and are underlined whenever the associated larger matrix is completed.

the same (n-1)-person problem as described at the beginning of the first step in the RC-rule, and so on.
From this dynamic interpretation of the RC-rule as well as the known formula (5.13) for the Shapley value in terms of orderings and the corresponding marginal contributions, we conclude that the RC-rule is related to the Shapley value. Their interrelationship is illustrated by the next example.

EXAMPLE 2.2. Consider again the simple bankruptcy problem with the estate $E = 400$ and $d_1 = 100$, $d_2 = 200$, $d_3 = 300$ as the involved claims. The corresponding 3-person bankruptcy game v is given by

$$v(1) = v(2) = 0, \quad v(3) = 100, \quad v(N) = 400,$$

$$v(12) = 100, \quad v(13) = 200, \quad v(23) = 300.$$

The marginal worth vectors $x^\theta(v)$, $\theta \in \Theta^3$, in the 3-person game v are listed in the fifth column of Table 6.2. Note that their calculations begin at the bottom of the estate, i.e., the payoffs according to any marginal worth vector reflect the increase from zero to the total estate E. The payoff vectors induced by the above dynamic interpretation of the RC-rule are listed in the columns 2-4 of Table 6.2. As opposed to the calculations for the determination of the marginal worth vectors, their calculations begin at the top of the estate, i.e., the payoffs according to the dynamic interpretation of the RC-rule reflect the decrease from the total estate E to the null estate. Consequently, any order of presenting the claims of the heirs at court is related to the ordering $\theta$ that describes the first (second, etc.) arrival of heir i (j, etc.) at court by means of $\theta(i) = n$ ($\theta(j) = n-1$, etc.). In other words, the vectors induced by all possible orders of presenting the claims of the heirs at court can be regarded as the marginal worth vectors with respect to the reversed orders. Due to the above dynamic interpretation of the RC-rule, we conclude that the Shapley value payoff vector $\phi(v)$ for the game v coincides with the division of the estate E by the RC-rule. This conclusion concerning the 3-person game v can be verified with the aid of the formula (5.13) because

$$\phi(v) = (3!)^{-1} \sum_{\theta\in\Theta^3} x^\theta(v) = \frac{1}{6}(400,700,1300) = RC(E;d).$$

Here the last equality was already derived from Figure 6.1.

| order of arrivals of the heirs | corresponding payoff to heir 1 | 2 | 3 | corresponding marginal worth vector $x^\theta(v)$ | ordering $\theta$ |
|---|---|---|---|---|---|
| (1 2 3) | 100 | 200 | 100 | (100,200,100) | (3 2 1) |
| (1 3 2) | 100 | 0 | 300 | (100, 0,300) | (2 3 1) |
| (2 1 3) | 100 | 200 | 100 | (100,200,100) | (3 1 2) |
| (2 3 1) | 0 | 200 | 200 | ( 0,200,200) | (1 3 2) |
| (3 1 2) | 100 | 0 | 300 | (100, 0,300) | (2 1 3) |
| (3 2 1) | 0 | 100 | 300 | ( 0,100,300) | (1 2 3) |

TABLE 6.2. The marginal worth vectors in comparison with the payoff vectors according to the dynamic interpretation of the RC-rule applied to the simple bankruptcy problem (E;d) of Example 2.2.

The reasonings as mentioned in Example 2.2 are also used to establish the general result that the recursive completion division rule for simple bankruptcy problems agrees with the Shapley value concept.

THEOREM 2.3. Let $(N;v_{E;d})$ be the bankruptcy game of (6.1) where the general bankruptcy problem (E;d) is simple. Then

$$RC(E;d) = \phi(v_{E;d}).$$

PROOF. We consider the dynamic interpretation of the RC-rule as well as the formula (5.13) for the Shapley value. Let $i \in N$ and $\sigma \in \Theta^n$. With respect to the given order $\sigma$ of presenting the claims of the heirs at court, the payoff $y_i^\sigma$ to heir i depends upon his claim $d_i$ and the amount what is left of the estate E after his predecessors j, $j \in P_i^\sigma$, have been paid off. The remaining part of the estate is equal to $E - d(P_i^\sigma)$ or zero, whichever is more. From this we deduce

$$\begin{aligned} y_i^\sigma &= d_i && \text{if} \quad E - d(P_i^\sigma) \ge d_i \\ &= E - d(P_i^\sigma) && \text{if } 0 \le E - d(P_i^\sigma) \le d_i \\ &= 0 && \text{if} \quad E - d(P_i^\sigma) \le 0. \end{aligned} \tag{6.2}$$

Let $\theta \in \Theta^n$ be the reverse order of $\sigma$, i.e.,

$$\theta(j) := n + 1 - \sigma(j) \qquad \text{for all } j \in N.$$

Then we get $P_i^\theta \cup \{i\} = \{j \in N |\ \theta(j) \le \theta(i)\}$

$$= \{j \in N |\ \sigma(j) \ge \sigma(i)\} = N - P_i^\sigma.$$

Put $w := v_{E;d}$. Further, for any $S \subset N$ we write $S^c$ instead of $N-S$. In view of $d_i \ge 0$ and the formula (6.1) for w, it is straightforward to verify that for all $S \subset N-\{i\}$ (6.3)

$$\begin{aligned} w(S^c) - w(S^c-\{i\}) &= d_i && \text{if } \quad E - d(S) \ge d_i \\ &= E - d(S) && \text{if } 0 \le E - d(S) \le d_i \\ &= 0 && \text{if } \quad E - d(S) \le 0. \end{aligned}$$

Now it follows immediately from (6.2) as well as (6.3) that

$$y_i^\sigma = w(S^c) - w(S^c-\{i\}) \qquad \text{applied to } S = P_i^\sigma.$$

Together with $N - P_i^\sigma = P_i^\theta \cup \{i\}$ and (5.10), this implies

$$y_i^\sigma = w(P_i^\theta \cup \{i\}) - w(P_i^\theta) = x_i^\theta(w) \tag{6.4}$$

where $\theta$ is the reverse order of $\sigma$.

Clearly, the equivalence $\theta \in \Theta^n$ iff $\sigma \in \Theta^n$ holds. Because $i \in N$, $\sigma \in \Theta^n$ are arbitrary, we conclude that for all $i \in N$

$$RC_i(E;d) = (n!)^{-1} \sum_{\sigma\in\Theta^n} y_i^\sigma = (n!)^{-1} \sum_{\theta\in\Theta^n} x_i^\theta(w) = \phi_i(w)$$

where the first equality follows from the dynamic interpretation of the RC-rule, the second equality from (6.4) and the third equality from the formula (5.13). Therefore, $RC(E;d) = \phi(v_{E;d})$ as was to be shown. □

### 3. The contested garment consistent division rule in comparison with the nucleolus

Aumann and Maschler (1985) interpreted the bankruptcy problem as the division problem of the estate of a died person among n creditors such that the n debts of the died person total at least as much as the estate. The approach of Aumann and Maschler is based on the idea of fixed debts and a variable estate. For example, they paid attention to the three bankruptcy problems with the three fixed debts $d_1 = 100$, $d_2 = 200$, $d_3 = 300$, and the variable estates $E = 100$, 200, and 300 respectively. These three bankruptcy problems were also considered in the Babylonian Talmud and the Mishna prescribed the divisions

$\frac{1}{3}(100,100,100)$, $(50,75,75)$, and $(50,100,150)$ respectively.

The first division can be seen as the egalitarian division of the estate $E = 100$, while the third division can be obtained by dividing the estate $E = 300$ in proportion to the three fixed debts. The recursive completion rule for simple bankruptcy problems can solely be applied to the third problem with the estate $E = 300$ and the corresponding RC-division coincides with the payoff vector $(50,100,150)$. Aumann and Maschler presented a common explanation for the three above-mentioned divisions in terms of the so-called contested garment principle.

The Mishna gives the next division rule for the problem of a contested garment (Baba Metzia, Babylonian Talmud):

> Two hold a garment; one claims it all, the other claims half of it. Then the former receives three quarters and the latter receives one quarter.

According to the contested garment principle, the lesser claimant concedes half the garment to the greater claimant and the remaining half is divided equally because both of them have an equal claim on the remaining half.

Aumann and Maschler applied the contested garment (CG-) principle to the two-person bankruptcy problem with the estate $E$ and the two claims $d_1$, $d_2$, as follows. In the first step, any creditor $i$ gets the amount $\max(0, E - d_j)$ that is conceded to creditor $i$ by the other creditor $j$ and in the second step, the remaining part of the estate is divided equally among the two creditors, i.e., both of them receive half of the remaining amount

$$E - \max(0, E - d_1) - \max(0, E - d_2).$$

The resulting division can be regarded as a function of the estate $E$ for the two fixed claims $d_1$ and $d_2$. The resulting payoffs $CG_i(E,d_1,d_2)$ to the creditors $i \in \{1,2\}$ are listed in Table 6.3 and are drawn in Figure 6.4 in case $d_1 \leq d_2$.

| | $0 \le E \le d_1$ | $d_1 \le E \le d_2$ | $d_2 \le E \le d_1 + d_2$ |
|---|---|---|---|
| $CG_1(E,d_1,d_2)$ | $\frac{1}{2}E$ | $\frac{1}{2}d_1$ | $\frac{1}{2}(d_1 + E - d_2)$ |
| $CG_2(E,d_1,d_2)$ | $\frac{1}{2}E$ | $E - \frac{1}{2}d_1$ | $\frac{1}{2}(d_2 + E - d_1)$ |

TABLE 6.3. The payoffs $CG_i(E,d_1,d_2)$ to creditors $i \in \{1,2\}$ according to the CG-principle for the two-person bankruptcy problem with the estate E and the two claims $d_1$, $d_2$, where $d_1 \le d_2$.

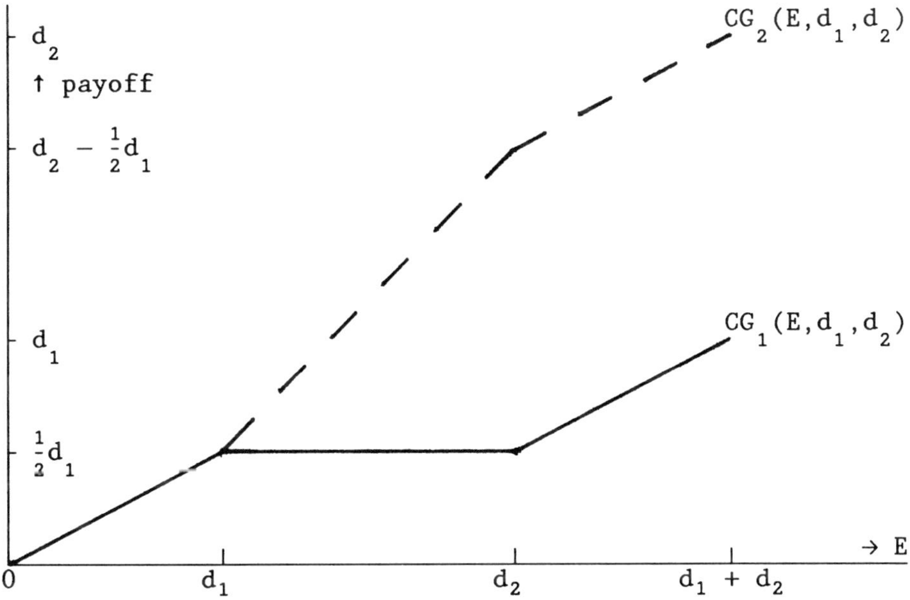


FIGURE 6.4. The solid (dotted respectively) curve represents the payoff $CG_i(E,d_1,d_2)$ to creditor $i = 1$ $(i = 2)$ as a function of E in case $d_1 \le d_2$.

The estate is shared equally between the two creditors when the estate is small enough. The egalitarian division of the total estate continues until both of them have received the payoff $\frac{1}{2}d_1$. Beyond this point, the greater claimant gets the entire additional estate until both of them have got all but $\frac{1}{2}d_1$ of their claims. Beyond that, the egalitarian method is used to divide the entire additional estate. From Figure 6.4 we deduce that each of the two resulting payoffs is a nondecreasing function of the estate.

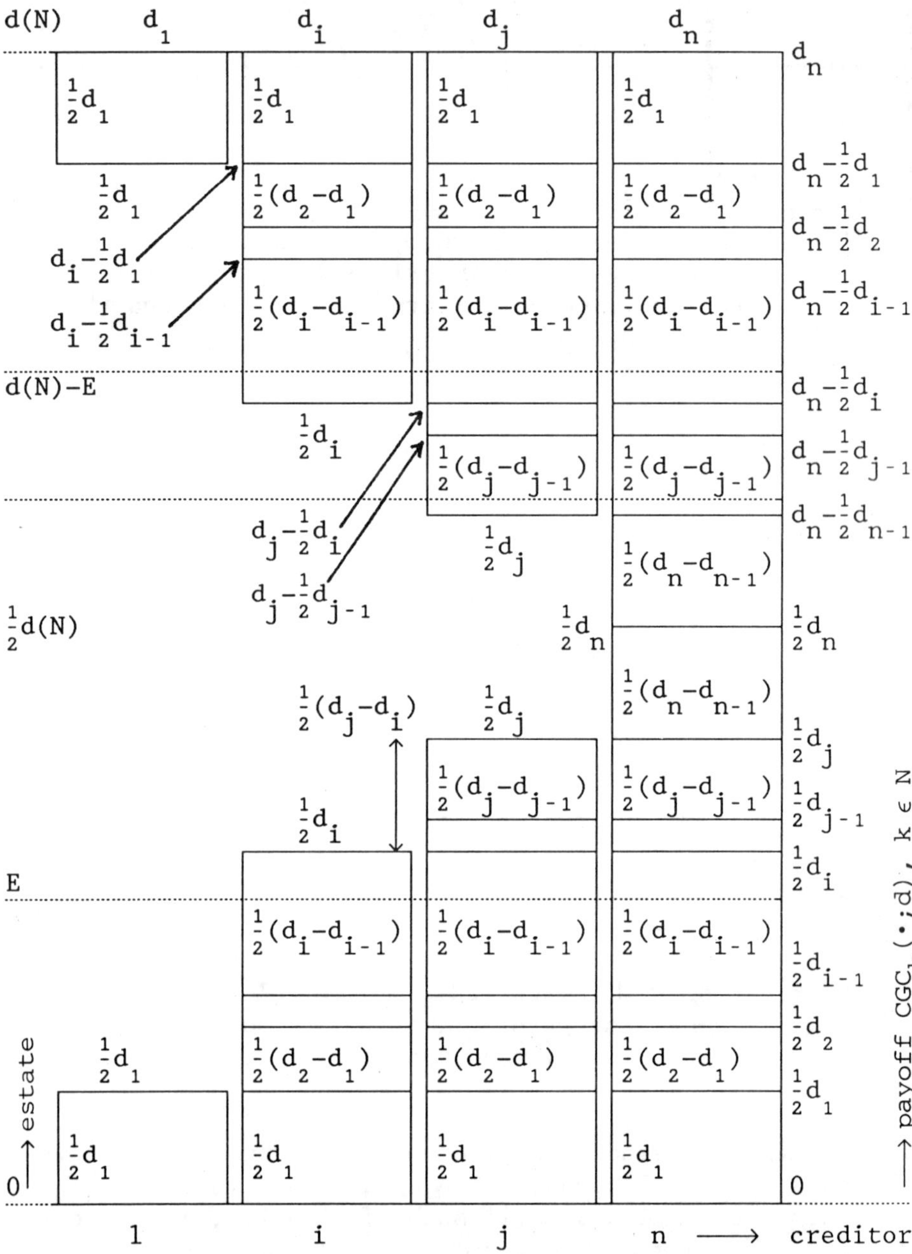

FIGURE 6.5. The division scheme according to the contested garment consistent (CGC-) rule. The payoff to any creditor is represented by the total area of the corresponding (two) rectangle(s). Here it is supposed that both
$1 \le i \le j \le n$ and $d_1 \le d_2 \le \cdots \le d_n$.

Now we describe a division rule for n-person bankruptcy problems which is a modification of the CG-principle for two-person bankruptcy problems. The division scheme of the rule in question is also drawn in Figure 6.5.
Consider the bankruptcy problem with a variable estate E and the fixed claims $d_1, d_2, \cdot\cdot, d_n$ of n creditors. Without loss of generality, these n claims may be ordered so that $d_1 \leq d_2 \leq \cdot\cdot \leq d_n$. The estate is shared equally between the n creditors if the estate is small enough. The egalitarian division of the estate continues until all of them have received the payoff $\frac{1}{2}d_1$ (which happens at the same moment). For the present, creditor 1 receives no further earnings and the entire additional estate is shared equally between the n-1 greater creditors. This continues until creditor 2 has received the payoff $\frac{1}{2}d_2$. For the moment, creditor 2 gets no further earnings and the entire additional estate is divided equally among the remaining n-2 creditors. Here this procedure continues until each creditor has received half his claim. At the half-way point E = 0.5d(N), there is an essential change in the division rule. If the variable estate does exceed 0.5d(N), then creditor n continues to receive the entire additional estate until he has got all but $\frac{1}{2}d_{n-1}$ of his claim $d_n$, i.e., the payoff $d_n - \frac{1}{2}d_{n-1}$. At the moment, creditor n shares the entire additional estate with creditor n-1 until both of them have received all but $\frac{1}{2}d_{n-2}$ of their claims (which happens at the same moment). At present, creditor n-2 also receives further earnings. This procedure continues until each creditor has got all but $\frac{1}{2}d_1$ of his claim. Eventually, the egalitarian method is used to divide the entire additional estate among the n creditors. The procedure stops at the point E = d(N) such that all claims are completely fulfilled. (6.5)

EXAMPLE 3.1. In order to illustrate the above rule, we look again at the bankruptcy problem with the three fixed claims $d_1 = 100$, $d_2 = 200$, $d_3 = 300$, and a variable estate E. Then the rule f of (6.5) agrees with the egalitarian division rule if the estate is small enough. Here the estate E is said to be small enough whenever $E \leq \frac{3}{2}d_1 = 150$. Therefore,

$f(E;d) = \frac{1}{3}(E, E, E)$ if $0 \leq E \leq 150$. In particular,

$f_i(150;d) = 50 = \frac{1}{2}d_1$ for all $i \in N$.

If $150 \le E \le 250$, then the entire additional estate $E - 150$ is shared equally between the two creditors 2 and 3. Thus,

$$f(E;d) = (50, \tfrac{1}{2}E - 25, \tfrac{1}{2}E - 25) \qquad \text{if } 150 \le E \le 250.$$

Particularly, $f_i(250;d) = 100 = \frac{1}{2}d_2$ for all $i \in \{2,3\}$. Next creditor 3 receives the entire additional estate $E - 250$ until his payoff $E - 150$ is equal to $d_3 - \frac{1}{2}d_2$. Due to this,

$$f(E;d) = (50, 100, E - 150) \qquad \text{if } 250 \le E \le 350.$$

As a consequence, $f(350;d) = (50, 100, 200)$. It is left to the reader to verify the following results:

$$f(E;d) = (50, \tfrac{1}{2}E - 75, \tfrac{1}{2}E + 25) \qquad \text{if } 350 \le E \le 450,$$

$$f(450;d) = (50, 150, 250),$$

$$f(E;d) = (\tfrac{1}{3}E - 100, \tfrac{1}{3}E, \tfrac{1}{3}E + 100) \qquad \text{if } 450 \le E \le 600.$$

With respect to the three estates $E = 100$, 200, and 300, the divisions according to the rule $f$ of (6.5) are equal to the divisions as prescribed by the Mishna.

Obviously, the division rule of (6.5) for n-person bankruptcy problems equals the CG-principle for 2-person bankruptcy problems whenever $n = 2$. As a matter of fact, the CG-principle is useful for a characterization of the division rule of (6.5). For that purpose, the notion of consistency is now treated.
Throughout the remainder of this section we tacitly suppose that a division rule $f$ satisfies the additional condition $f_i(E;d) \le d_i$ for all $i \in N$ which condition expresses that the division for any general bankruptcy problem $(E;d)$ is bounded above by the claims of the creditors.
Consider a division rule $f$ and let $(E;d)$ be any bankruptcy problem. The CG-principle can be applied to the two-person bankruptcy problem with the two claims $d_i$ of creditor $i$ and $d_j$ of creditor $j$ respectively in order to divide the total amount $f_i(E;d) + f_j(E;d)$ that is apportioned to their two-person coalition $\{i,j\}$ according to the involved rule $f$. The division rule $f$ is called CG-consistent if the division according to the CG-principle for the above two-creditor problem is given by the payoff $f_i(E;d)$ to creditor $i$ and the payoff $f_j(E;d)$ to creditor $j$.

That is, a division rule f is said to be <u>CG-consistent</u> if for all bankruptcy problems (E;d) and all $i,j \in N$, $i \neq j$,

$$CG\big(f_i(E;d) + f_j(E;d),\ d_i,\ d_j\big) = \big(f_i(E;d),\ f_j(E;d)\big). \qquad (6.6)$$

Roughly speaking, a division rule f is CG-consistent if and only if the rule f agrees with the CG-principle applied to any two-creditor problem with their two fixed claims and their total payoff according to the involved division rule.

THEOREM 3.2. The division rule of (6.5) is the unique CG-consistent division rule.

PROOF. (a) We first prove the uniqueness part. Assume that there exist at least two CG-consistent division rules g, h. Then there exists a bankruptcy problem (E;d) such that $g(E;d) \neq h(E;d)$. Put $x := g(E;d)$ and $y := h(E;d)$. We have $x(N) = E = y(N)$ by the efficiency of the two division rules g and h. From $x \neq y$ and $x(N) = y(N)$ we deduce that there exist $i,j \in N$, $i \neq j$, with $x_i > y_i$ as well as $x_j < y_j$. Now we conclude from the consistency condition (6.6) applied to the two division rules g and h that

$$CG_i(x_i + x_j,\ d_i,\ d_j) = x_i, \quad CG_j(x_i + x_j,\ d_i,\ d_j) = x_j,$$

$$CG_i(y_i + y_j,\ d_i,\ d_j) = y_i, \quad CG_j(y_i + y_j,\ d_i,\ d_j) = y_j.$$

Since each of the two payoffs according to the CG-principle is a nondecreasing function of the estate, we obtain that in case $x_i + x_j \leq y_i + y_j$

$$CG_i(x_i + x_j,\ d_i,\ d_j) \leq CG_i(y_i + y_j,\ d_i,\ d_j) \quad \text{and thus,}$$

$$x_i = CG_i(x_i + x_j,\ d_i,\ d_j) \leq CG_i(y_i + y_j,\ d_i,\ d_j) = y_i$$

where the obtained result $x_i \leq y_i$ is in contradiction with the strict inequality $x_i > y_i$. In case $x_i + x_j \geq y_i + y_j$, the so obtained result $x_j \geq y_j$ is contradicted by $x_j < y_j$. In any case we have derived a contradiction. Hence, there is at most one CG-consistent division rule. This proves the uniqueness part.

(b) Next we prove the CG-consistency of the rule of (6.5). Let (E;d) be the bankruptcy problem with the estate E and

the claims $d_1, d_2, \cdots, d_n$, satisfying $d_1 \le d_2 \le \cdots \le d_n$. Let $i,j \in N$, $i \ne j$, and suppose $d_i \le d_j$ holds. We verify the consistency condition (6.6) for the rule f of (6.5) by using Table 6.3 and Figure 6.5. Put $E^* := f_i(E;d) + f_j(E;d)$. Due to the efficiency of the CG-principle, it suffices to show $CG_i(E^*,d_i,d_j) = f_i(E;d)$. We distinguish three cases.
Case one. Suppose $0 \le f_j(E;d) < \frac{1}{2}d_i$ (see the lower dotted line in Figure 6.5). The construction of the rule f yields $f_i(E;d) = f_j(E;d)$. Therefore, $E^* = 2f_i(E;d)$ and $0 \le E^* < d_i$ as well. Together with Table 6.3, this implies the equality $CG_i(E^*,d_i,d_j) = \frac{1}{2}E^* = f_i(E;d)$.
Case two. Suppose $\frac{1}{2}d_i \le f_j(E;d) \le d_j - \frac{1}{2}d_i$ (see the middle dotted line in Figure 6.5). The construction of the rule f yields $f_i(E;d) = \frac{1}{2}d_i$. Thus, $d_i \le E^* \le d_j$ and the equality $CG_i(E^*,d_i,d_j) = \frac{1}{2}d_i = f_i(E;d)$ is deduced from Table 6.3.
Case three. Suppose $d_j - \frac{1}{2}d_i < f_j(E;d) \le d_j$ (see the upper dotted line in Figure 6.5). The construction of the rule f yields $f_j(E;d) - f_i(E;d) = d_j - d_i$ and $\frac{1}{2}d_i \le f_i(E;d) \le d_i$. Hence, $d_j < E^* \le d_i + d_j$ and it follows from Table 6.3 that

$$CG_i(E^*,d_i,d_j) = \tfrac{1}{2}(d_i + E^* - d_j)$$
$$= \tfrac{1}{2}\big(E^* - f_j(E;d) + f_i(E;d)\big) = f_i(E;d).$$

In any case we have shown $CG_i(E^*,d_i,d_j) = f_i(E;d)$ and thus, (6.6) holds. This completes the proof of the CG-consistency of the division rule f of (6.5). □

In view of Theorem 3.2, the division rule of (6.5) is called the <u>contested garment consistent</u> (CGC-) <u>rule</u>. For the three bankruptcy problems with the three fixed claims $d_1 = 100$, $d_2 = 200$, $d_3 = 300$, and the variable estates $E = 100$, 200, and 300 respectively, the divisions of the three estates according to the CGC-rule coincide with the divisions as prescribed by the Mishna, while it is left to the reader to check that these divisions are also equal to the nucleoli of the corresponding bankruptcy games. As a matter of fact, the contested garment consistent division rule agrees with the nucleolus concept.

THEOREM 3.3. Let $(N;v_{E;d})$ be the bankruptcy game of (6.1). Then

$$CGC(E;d) = \eta(v_{E;d}).$$

Because of Theorem V.7.4, the statement in Theorem 3.3 is equivalent to the statement that the CGC-rule agrees with the prekernel concept. Hence, the proof of Theorem 3.3 is mainly based on the prekernel and particularly, a so-called reduced game property for the prekernel. Since the theory concerning the reduced games will not be considered in this work, we omit the proof of Theorem 3.3.

## 4. The adjusted proportional division rule in comparison with the $\tau$-value

In Section IV.2 we presented three separable cost allocation methods in the water resources field. The three methods meet the principle that no user in a multipurpose water resource project should be charged less than the separable cost of including the user in the joint project. Curiel, Maschler and Tijs (1987) introduced a division rule which meets the analogous principle that no creditor in a bankruptcy problem should be awarded less than his minimum right. The <u>minimum right</u> $m_i$ <u>of creditor</u> i in a bankruptcy problem (E;d) is defined as the part of the estate that is not claimed by the complementary coalition N-{i}, i.e., for all $i \in N$

$$m_i := \max\big(0,\ E - d(N-\{i\})\big) = v_{E;d}(\{i\}). \qquad (6.7)$$

Notice that $0 \le m_i \le d_i$ for all $i \in N$ and furthermore, it follows from the convexity of the bankruptcy game $v_{E;d}$ that

$$m(N) = \sum_{j\in N} m_j = \sum_{j\in N} v_{E;d}(\{j\}) \le v_{E;d}(N) = E.$$

Given that each creditor has received his minimum right, the remaining part of the estate is divided in proportion to the adjusted claims of the creditors. Here the adjusted claims are obtained as follows. In the first step, the original claim of any creditor is lowered by his minimum right already received and in the second step, the lowered claims $d_i - m_i$, $i \in N$, are truncated with respect to the remaining estate $E - m(N)$. As such, we arrive at the simple bankruptcy problem with the estate $E - m(N)$ and the claims

$$d_i^* := \min\left(d_i - m_i,\ E - m(N)\right) \qquad \text{for all } i \in N. \tag{6.8}$$

This induced bankruptcy problem is solved by means of the proportional division rule which divides the remaining estate $E - m(N)$ among the creditors proportional to their adjusted claims $d_i^*$, $i \in N$.

Due to the above reasoning, the resulting division rule is called the <u>adjusted proportional</u> (AP-) <u>rule</u> and the general formula for its division in a bankruptcy problem $(E;d)$ is as follows: for all $i \in N$

$$\begin{aligned} AP_i(E;d) &= m_i && \text{if } E = m(N) \\ &= m_i + \left(E - m(N)\right)\left(d^*(N)\right)^{-1} d_i^* && \text{if } E > m(N) \end{aligned} \tag{6.9}$$

where $m_i$ and $d_i^*$ respectively are given by (6.7) and (6.8).

EXAMPLE 4.1. Consider once again the bankruptcy problem $(E;d)$ of Example 2.2 and the corresponding bankruptcy game. That is, $E = 400$ and $d_1 = 100$, $d_2 = 200$, $d_3 = 300$ as well. So, we obtain that the minimum right vector $m = (0,0,100)$, the remaining estate $E - m(N) = 300$ and the adjusted claims are determined by means of $d_i^* = d_i - m_i$ for all $i \in N$. Now the proportional division of the estate 300 in the induced simple bankruptcy problem with the adjusted claims vector $d^* = (100,200,200)$ is equal to $0.6d^*$. We conclude that the division according to the AP-rule is given by the vector

$AP(E;d) = m + \frac{3}{5}d^* = (60,120,220)$. We recall that

$RC(E;d) = \frac{1}{3}(200,350,650)$ and

$CGC(E;d) = (50,125,225)$.

We also determine the $\tau$-value of the corresponding 3-person bankruptcy game v. By the formulas (3.1)-(3.3), we get that

$$b^v = (100,200,300), \quad g^v(\{1\}) = 100,$$
$$g^v(S) = 200 \qquad \text{for all } S \neq \{1\},\emptyset,$$
$$\lambda_i^v = g^v(\{i\}) \qquad \text{for all } i \in N.$$

Hence, the division of the joint concession amount $g^v(N)$ is proportional to the concession vector $\lambda^v = (100,200,200)$. We conclude that the $\tau$-value payoff vector for the game v is given by $\tau(v) = b^v - 0.4\lambda^v = (60,120,220)$. Observe that the $\tau$-value payoff vector coincides with the AP-division.

Similar reasonings as mentioned in the above example are also used to establish the general result that the adjusted proportional division rule agrees with the $\tau$-value concept.

THEOREM 4.2. Let $(N;v_{E;d})$ be the bankruptcy game of (6.1). Then

$$AP(E;d) = \tau(v_{E;d}).$$

PROOF. Put $w := v_{E;d}$. As noted in Section V.4, the $\tau$-value of the convex game w represents some efficient compromise between the upper bound $b^w$ and the lower bound $b^w - \lambda^w$ for the core of the game where $b_i^w - \lambda_i^w = w(\{i\})$ for all $i \in N$. In view of the relevant formula (6.9), the AP-division for the bankruptcy problem (E;d) represents the identically efficient compromise between the minimum right vector m and the vector $m + d^*$ where $d^*$ is the adjusted claims vector. Note that $m_i = w(\{i\})$ for all $i \in N$ by the formula (6.7). Now it follows from the geometric point of view that the $\tau$-value $\tau(w)$ coincides with the division AP(E;d) whenever the two corresponding upper bounds $b^w$ and $m + d^*$ coincide. Thus, it suffices to show the vector equality $m + d^* = b^w$. Let $i \in N$. By the formulas (3.1), (6.1) and (6.8), we have

$$b_i^w = w(N) - w(N-\{i\}) = E - \max\left(0,\ E - d_i\right) = \min\left(d_i, E\right),$$

$$m_i + d_i^* = \min\left(d_i,\ E - m(N-\{i\})\right) \qquad \text{and}$$

$$E - m(N-\{i\}) = w(N) - \sum_{j\in N-\{i\}} w(\{j\})$$

$$\geq w(N) - w(N-\{i\}) = b_i^w$$

where the inequality is due to the convexity of the game w.

In case $d_i \leq E$, then we obtain

$$d_i = b_i^w \leq E - m(N-\{i\}) \quad \text{and hence,} \quad m_i + d_i^* = d_i = b_i^w.$$

In case $d_i > E$, then we obtain that for all $j \in N-\{i\}$

$$E - d(N-\{j\}) \leq E - d_i < 0 \quad \text{and so,} \quad m_j = w(\{j\}) = 0$$

$$\text{and as a consequence,} \qquad m_i + d_i^* = \min(d_i, E) = b_i^w.$$

In any case we conclude that $m + d^* = b^w$. □

## 5. Axiomatic characterizations of division rules in terms of self-duality

In Section 3 we described the CGC-division rule in terms of awards. However, the CGC-rule can also be viewed from a different angle when the variable estate passes the half-way point 0.5d(N). If the estate E exceeds 0.5d(N), then each creditor receives at least half his claim and so, instead of thinking in terms of his award, each creditor thinks in terms of his loss with respect to his own claim. The total loss d(N) - E is shared equally between the n creditors when this loss is small enough. The egalitarian division of the total loss continues until all of them have lost the amount $\frac{1}{2}d_1$ (which happens at the same moment). Beyond this point, creditor 1 is free from any further loss and the entire additional loss is shared equally between the n-1 greater creditors. This continues until creditor 2 has lost the amount $\frac{1}{2}d_2$. Beyond that, creditor 2 stops losing and the entire additional loss is divided equally among the remaining n-2 creditors. This procedure continues until each creditor has lost half his claim which happens at the half-way point E = 0.5d(N).

We emphasize the significant change in the CGC-rule at the half-way point E = 0.5d(N). In case the variable estate is at most (at least respectively) 0.5d(N), then each creditor receives at most (at least) half his claim and thus, each creditor thinks in terms of his award (loss with respect to his claim). Hence, awards and losses are treated dually and moreover, an award of $x_i$ to creditor i can be interpreted as a loss of $d_i - x_i$ to creditor i. Aumann and Maschler (1985) introduced the notion of self-duality to express that awards and losses are treated in the same way.

DEFINITION 5.1. A division rule f is said to be <u>self-dual</u> if for all bankruptcy problems (E;d)

$$f(E;d) = d - f(d(N)-E;d). \tag{6.10}$$

The self-duality condition (6.10) for a division rule requires that the resulting award to any creditor in the original bankruptcy problem (E;d) is equal to his loss with respect to his claim in the interrelated bankruptcy problem (d(N)-E;d) in which the new estate d(N) - E represents the total amount of unfulfilled claims.
The self-duality of the known CGC-division rule is a direct consequence of its construction, but is also visualized in Figure 6.5 by the symmetry of the CGC-division scheme with respect to the half-way point $E = 0.5d(N)$. Evidently, the equivalence $E \leq 0.5d(N)$ iff $d(N) - E \geq 0.5d(N)$ holds. For example, in case the estate E is at most 0.5d(N), then the award $CGC_i(E;d)$ to creditor i in the original bankruptcy problem (E;d) is represented in Figure 6.5 by the area of the part of the corresponding rectangle which lies below the lower dotted line, whereas the area of the part of the second rectangle which lies above the upper dotted line represents the loss $d_i - CGC_i(d(N)-E;d)$ of creditor i with respect to his claim in the interrelated bankruptcy problem (d(N)-E;d). In general, both areas are equal and therefore, the CGC-division rule is self-dual.
The next theorem states that the RC-division rule possesses the self-duality property restricted to the class of simple bankruptcy problems. The proof of the theorem is based on the known fact that the RC-division rule agrees with the Shapley value concept. In a similar way we establish that the AP-division rule is self-dual as well.

THEOREM 5.2. Let (E;d) be a bankruptcy problem such that $d_i \leq E \leq d(N-\{i\})$ for all $i \in N$. Then

$$RC(E;d) = d - RC(d(N)-E;d).$$

PROOF. We denote by (N;v) and (N;w) respectively the two bankruptcy games associated with the bankruptcy problems (E;d) and (d(N)-E;d). Further, for any $S \subset N$ we write $S^c$ instead of N-S. By the formula (6.1) applied to the bankruptcy games v and w, we have that for all $S \subset N$

$v(S) = \max(0,\ E - d(S^c))$, $w(S) = \max(0,\ d(S) - E)$ and so,

$$v(S) = w(S^c) + E - d(S^c) \qquad \text{for all } S \subset N. \tag{6.11}$$

From (6.11) we obtain that for all $i \in N$ and all $S \subset N-\{i\}$

$$v(S \cup \{i\}) - v(S) = w(S^c-\{i\}) - w(S^c) + d_i$$

while $\gamma_n(S) = (n!)^{-1}|S|!(n-|S|-1)! = \gamma_n(S^c-\{i\})$. Together with the formula (2.3), this implies that for all $i \in N$

$$\phi_i(v) = \sum_{S\subset N-\{i\}} \gamma_n(S)\big(v(S \cup \{i\}) - v(S)\big)$$

$$= \sum_{S\subset N-\{i\}} \gamma_n(S^c-\{i\})\big(w(S^c-\{i\}) - w(S^c) + d_i\big)$$

$$= \sum_{T\subset N-\{i\}} \gamma_n(T)\big(w(T) - w(T \cup \{i\}) + d_i\big)$$

$$= \sum_{T\subset N-\{i\}} \gamma_n(T)d_i - \phi_i(w) = d_i\cdot 1 - \phi_i(w) = d_i - \phi_i(w).$$

Thus, the vector equality $\phi(v) = d - \phi(w)$ holds. From this vector equality and Theorem 2.3, we conclude that

$$RC(E;d) = \phi(v) = d - \phi(w) = d - RC(d(N)-E;d). \qquad \square$$

THEOREM 5.3. $AP(E;d) = d - AP(d(N)-E;d)$ for all bankruptcy problems $(E;d)$.

PROOF. We use the same notation as mentioned at the beginning of the previous proof. From the two formulas (3.1) and (6.11) we obtain that for all $i \in N$

$$\begin{gathered} b_i^w = w(N) - w(N-\{i\}) = d(N) - E - w(N-\{i\}) = d_i - v(\{i\}), \\ w(\{i\}) = v(N-\{i\}) - E + d_i = v(N-\{i\}) - v(N) + d_i = d_i - b_i^v. \end{gathered} \tag{6.12}$$

Further, the bankruptcy games v and w are convex. As noted in Section V.4, the $\tau$-value of any convex game u represents some efficient compromise between the upper bound $b^u$ and the lower bound $b^u - \lambda^u$ for the core of the game such that $b_i^u - \lambda_i^u = u(\{i\})$ for all $i \in N$. In view of this geometric interpretation of the $\tau$-value for convex games and (6.12), it follows that the vector equality $\tau(w) = d - \tau(v)$ holds. Together with Theorem 4.2, this vector equality implies

$$AP(E;d) = \tau(v) = d - \tau(w) = d - AP(d(N)-E;d). \qquad \square$$

So far, we showed that the self-duality condition (6.10) is satisfied by the CGC-, RC- as well as the AP-division rule. The remainder of this section is devoted to an axiomatic characterization of the CGC- and the AP-rule respectively by means of the self-duality and an additional property.

We first treat an additional property in terms of the constrained equal awards. A division of the estate is said to be a <u>constrained equal award division</u> for a bankruptcy problem (E;d) if each creditor receives the same award on the understanding that any creditor gets at most his claim, i.e., there exists a real number $\alpha$ such that any creditor i receives the award $\min(d_i, \alpha)$. The next lemma states that any bankruptcy problem generates a unique constrained equal award division of the total estate. Notice that the unique constrained equal award division for a bankruptcy problem (E;d) with $E = d(N)$ is given by the claims vector d itself.

LEMMA 5.4. Let (E;d) be a bankruptcy problem with $d(N) > E$. Then there exists a unique real number $\alpha$ such that

$$\sum_{j\in N} \min(d_j, \alpha) = E.$$

PROOF. Define the real-valued functions $h^i\colon \mathbb{R} \to \mathbb{R}$, $i \in N$, and their sum $h\colon \mathbb{R} \to \mathbb{R}$ by

$$h^i(\alpha) := \min(d_i, \alpha) \qquad \text{for all } \alpha \in \mathbb{R},$$

$$h(\alpha) := \sum_{j\in N} h^j(\alpha) = \sum_{j\in N} \min(d_j, \alpha) \qquad \text{for all } \alpha \in \mathbb{R}.$$

The functions $h^i$, $i \in N$, are continuous and nondecreasing on $\mathbb{R}$, while the function $h^k$ is even increasing on $[0,d_k]$. Here $k \in N$ is chosen such that $d_i \le d_k$ for all $i \in N$. It follows that the sum h is a continuous increasing function on $[0,d_k]$ satisfying $h([0,d_k]) = [0,d(N)]$. In view of this and $0 \le E \le d(N)$, we conclude that there exists a unique real number $\alpha \in [0,d_k]$ with $h(\alpha) = E$. Furthermore, we have

$$h(\alpha) = n\alpha \quad < 0 \qquad \text{if } \alpha < 0,$$

$$h(\alpha) = d(N) > E \qquad \text{if } \alpha > d_k.$$

Hence, there exists a unique $\alpha \in \mathbb{R}$ such that $h(\alpha) = E$. □

A division rule f is said to possess the <u>constrained equal award property</u> if, for all bankruptcy problems (E;d) with $E \le 0.5d(N)$, the resulting efficient payoff vector f(E;d) coincides with the unique constrained equal award division for the corresponding bankruptcy problem (E;0.5d) with the original estate E and the reduced claims vector 0.5d, i.e., there exists a real number $\alpha$ such that

$$f_i(E;d) = \min\left(\tfrac{1}{2}d_i, \alpha\right) \qquad \text{for all } i \in N.$$

It is a direct consequence of its construction that the CGC-division rule possesses the constrained equal award property, but the property can also be deduced from its division scheme (see the lower dotted line in Figure 6.5). As a matter of fact, the CGC-rule is completely axiomatized by this property and the self-duality property.

THEOREM 5.5. The CGC-rule is the unique division rule with the self-duality property and the constrained equal award property.

PROOF. It remains to prove the uniqueness part. Suppose that both g and h are division rules with the two mentioned properties. We show that g(E;d) = h(E;d) for all bankruptcy problems (E;d). If $E \le 0.5d(N)$, then it follows from the constrained equal award property for g and h respectively that the division g(E;d) as well as the division h(E;d) coincide with the unique constrained equal award division for the corresponding bankruptcy problem (E;0.5d) and thus, g(E;d) = h(E;d) whenever $E \le 0.5d(N)$. In case $E \ge 0.5d(N)$, then $d(N) - E \le 0.5d(N)$ and hence,

$$g(E;d) = d - g(d(N)-E;d) = d - h(d(N)-E;d) = h(E;d)$$

where the first and third equality follow from the self-duality properties for g and h, while the second equality is due to the constrained equal award property for g and h as well. We obtain that g(E;d) = h(E;d) for all bankruptcy problems (E;d). This proves the uniqueness part. □

Finally, an axiomatization of the AP-rule can be based on another additional property. A division rule f is said to possess the <u>truncated claim proportionality property</u> if, for all bankruptcy problems (E;d) with $E \le 0.5d(N)$, the resulting efficient payoff vector f(E;d) is proportional to the truncated claims vector $d' \in \mathbb{R}^n$ defined by

$$d'_i := d_i \quad \text{if} \quad d_i \le E$$
$$= E \quad \text{if } E \le d_i \le d(N) - E$$
$$= 2E - d(N-\{i\}) \quad \text{if} \quad d_i \ge d(N) - E.$$

The claims vector d' is obtained by truncating the original claims vector d with respect to the total estate E on the understanding that the given claims are lowered by at most the amount d(N) - 2E. Obviously, the vector equality d' = d holds whenever the bankruptcy problem (E;d) is simple.

EXAMPLE 5.6. Consider once again the bankruptcy problem with the three fixed claims $d_1 = 100$, $d_2 = 200$, $d_3 = 300$, and a variable estate E where $0 < E \le 300$. By (6.7)-(6.9), we get that $m_i = 0$ for all $i \in N$ and hence, the AP-division is proportional to the adjusted claims vector $d^*$ satisfying $d^*_i = \min(d_i, E)$ for all $i \in N$. Now it is straightforward to verify the equality $d^* = d'$. We conclude that the division of the estate according to the AP-rule is in proportion to the truncated claims vector d'. The next proposition states that this result holds in general.

PROPOSITION 5.7. The AP-division rule possesses the truncated claim proportionality property.

PROOF. Let (E;d) be a bankruptcy problem with $E \le 0.5d(N)$. Choose any player $k \in N$ satisfying $d_i \le d_k$ for all $i \in N$. We distinguish the two cases $E \le d(N-\{k\})$ and $E \ge d(N-\{k\})$. Case one. Suppose $E \le d(N-\{k\})$. We have that for all $i \in N$

$$E \le d(N-\{k\}) \le d(N-\{i\}) \quad \text{and hence,}$$
$$d'_i = \min(d_i, E) \quad \text{as well as}$$
$$m_i = 0 \quad \text{by using (6.7).}$$

Together with (6.8), this implies that for all $i \in N$

$$d^*_i = \min(d_i - m_i, E - m(N)) = \min(d_i, E) = d'_i.$$

We obtain that $m_i = 0$ and $d^*_i = d'_i$ for all $i \in N$. In view of (6.9), the AP-division represents some efficient compromise between the minimum right vector m and the vector $m + d^*$. Now it follows that the division AP(E;d) is proportional to the vector d'.

Case two. Suppose $E \geq d(N-\{k\})$. We have that for all $i \neq k$

$$d_i \leq d(N-\{k\}) \leq E \leq d(N) - E \quad \text{and hence,}$$

$$d'_k = 2E - d(N-\{k\}), \qquad m_k = E - d(N-\{k\}),$$

$$d'_i = d_i, \qquad m_i = 0 \qquad \text{for all } i \in N-\{k\}.$$

It follows that $E - m(N) = E - m_k = d(N-\{k\})$ and so, we get

$$d^*_k = \min\big(d_k - m_k,\ E - m(N)\big)$$

$$= \min\big(d(N) - E,\ d(N-\{k\})\big) = d(N-\{k\}),$$

$$d^*_i = \min\big(d_i - m_i,\ E - m(N)\big)$$

$$= \min\big(d_i,\ d(N-\{k\})\big) = d_i \qquad \text{for all } i \in N-\{k\}. \quad \text{Thus,}$$

$$(m + \tfrac{1}{2}d^*)(N) = m(N) + \tfrac{1}{2}d^*(N) = m_k + d(N-\{k\}) = E.$$

Because the vector $m + 0.5d^*$ is the efficient compromise between the minimum right vector $m$ and the vector $m + d^*$, we conclude that $AP(E;d) = m + 0.5d^*$. We also deduce that

$$m_k + \tfrac{1}{2}d^*_k = E - \tfrac{1}{2}d(N-\{k\}) = \tfrac{1}{2}d'_k,$$

$$m_i + \tfrac{1}{2}d^*_i = \tfrac{1}{2}d_i = \tfrac{1}{2}d'_i \qquad \text{for all } i \in N-\{k\}.$$

We have proved the equality $AP(E;d) = m + 0.5d^* = 0.5d'$. In any case, the division $AP(E;d)$ is proportional to the truncated claims vector $d'$ as was to be shown. □

THEOREM 5.8. The AP-rule is the unique division rule with the self-duality property and the truncated claim proportionality property.

PROOF. In view of both Theorem 5.3 and Proposition 5.7, it remains to prove the uniqueness part. Suppose that $g$ and $h$ are division rules with the two mentioned properties. We show that $g(E;d) = h(E;d)$ for any bankruptcy problem $(E;d)$. If $E \leq 0.5d(N)$, then it follows from the truncated claim proportionality property for $g$ and $h$ respectively that $g(E;d) = \alpha d'$ and $h(E;d) = \beta d'$ where $\alpha, \beta \in \mathbb{R}$. The efficiency property for $g$ and $h$ yields $\alpha d'(N) = E = \beta d'(N)$. Therefore, either $\alpha = \beta$ or $d'(N) = 0$ which last equality is equivalent to $d'_i = 0$ for all $i \in N$. We obtain that

$$g(E;d) = \alpha d' = \beta d' = h(E;d) \qquad \text{whenever } E \le 0.5d(N).$$

In case $E \ge 0.5d(N)$, then $d(N) - E \le 0.5d(N)$ and hence,

$$g(E;d) = d - g(d(N)-E;d) = d - h(d(N)-E;d) = h(E;d)$$

where the first and third equality are due to the self-duality properties for g and h, while the second equality follows from the truncated claim proportionality property for g and h as well. We obtain that $g(E;d) = h(E;d)$ for all bankruptcy problems $(E;d)$. This proves the uniqueness part. □

The above axiomatic characterization of the CGC-rule is due to Aumann and Maschler (1985), while Curiel, Maschler and Tijs (1987) proved the above axiomatization of the AP-rule. It is an open problem to characterize the RC-rule by means of the self-duality and one or more additional properties. We conclude this chapter with a treatment of the problem of rabbi Ibn Ezra as mentioned at the beginning of Section I.6. The involved problem deals with the division of Jacob's estate among his four sons Reuben, Simeon, Levi and Judah.

EXAMPLE 5.2. We consider the bankruptcy problem $(E;d)$ with the four claims $d_1 = 30$, $d_2 = 40$, $d_3 = 60$, $d_4 = 120$ and the estate $E = 120$. Since the bankruptcy problem is simple such that $E \le 0.5d(N)$, the truncated claims vector $d'$ is equal to the original claims vector d and hence, the AP-division of the estate is proportional to the claims vector d, i.e.,

$$AP(E;d) = 0.48(30,40,60,120)$$
$$= (14.4,\ 19.2,\ 28.8,\ 57.6).$$

According to the CGC-division rule, each creditor receives half his claim with the exception of the greatest claimant and thus, $CGC(E;d) = (15,20,30,55)$. Further, we determine the RC-division of the estate with the aid of the Shapley value for the associated 4-person bankruptcy game v. The game v itself is given by

$$v(14) = 20, \qquad v(24) = 30, \qquad v(34) = 50,$$
$$v(124) = 60, \qquad v(134) = 80, \qquad v(234) = 90,$$
$$v(N) = 120, \qquad v(S) = 0 \qquad \text{for all other } S \subset N.$$

In turn, the computation of the Shapley value is based on the marginal worth vectors in the game v which are given by

$$( 0, 0, 0,120), \quad ( 0, 0,60,60), \quad ( 0,40,60,20),$$
$$( 0,40, 0, 80), \quad (20,40,60, 0), \quad (30, 0, 0,90),$$
$$(30, 0,60, 30), \quad (30,30,60, 0), \quad (30,40, 0,50),$$
$$(30,40,50, 0).$$

The multiplicity of any marginal worth vector is equal to 2 except for the point (0,0,0,120) which has multiplicity 6. By the formula (5.13), we obtain

$$RC(E;d) = \phi(v) = \frac{1}{24}(340,460,700,1380)$$

$$= (14.17,\ 19.17,\ 29.17,\ 57.5).$$

We observe that the CGC-division is preferred to both the AP- and the RC-division by all but the greatest claimant who prefers the payoff according to the AP-rule.

The general bankruptcy problem forms part of a much wider class of problems in distributive justice. The class consists of problems of how to distribute a fixed resource or cost among individuals according to their fixed claims on the resource or their potential liabilities for the cost. For instance, the taxation problem can be interpreted as the problem of dividing a tax among citizens according to their ability to pay. In the context of taxation, the fairness of division rules is studied in Young (1987) on the basis of a so-called consistency property for division rules. The consistency property in question is formulated as the consistency condition (6.6) for division rules in which the CG-principle is replaced by the division rule itself. Young (1987) also illustrated the application of his criteria of a satisfactory taxation rule to the three bankruptcy problems which were considered in the Babylonian Talmud.

# CHAPTER VII

# k-CONVEX GAMES AND SOLUTION CONCEPTS

## 1. k-Convex games

In Section V.3 we presented three characterizations of convex games in terms of the marginal worth vectors and the core. One of those three characterizations expresses that an n-person game v is convex if and only if all n! marginal worth vectors $x^{\theta}(v)$, $\theta \in \Theta^n$, belong to the core C(v) of the game. According to Theorem III.5.4, a slightly analogous characterization is also available for 1-convex games which were introduced in Section III.5. As a matter of fact, an n-person game v is 1-convex if and only if all n adjusted efficient upper vectors $b^v - g^v(N)e^i$, $i \in N$, belong to the core C(v) of the game.
In addition, the resemblance of 1-convexity to convexity can be described in terms of suitably chosen core catchers. In view of Theorem V.3.6, the convex hull of the set of all marginal worth vectors in a game is always a core catcher and as such, it coincides with the core if and only if the game is convex. Due to the next theorem, the convex hull of the set of all adjusted efficient upper vectors in a game is also a core catcher and as such, it coincides with the core merely for 1-convex games.

THEOREM 1.1.
(i) $C(v) \subset \text{conv}\,\{b^v - g^v(N)e^i \mid i \in N\}$ for all $v \in G^n$.
(ii) $C(v) = \text{conv}\,\{b^v - g^v(N)e^i \mid i \in N\}$ iff $v \in C_1^n$.

PROOF. Let $v \in G^n$. For any $i \in N$ we put $x^{(i)} := b - g(N)e^i$.
(i) We show that $C(v) \neq \emptyset$ implies $C(v) \subset \text{conv}\{x^{(i)} \mid i \in N\}$. Suppose $C(v) \neq \emptyset$. By using Lemma III.1.2(ii), we have that $g(S) \geq 0$ for all $S \subset N$. In case $g(N) = 0$, then $C(v) = \{b\}$ by Proposition III.3.4 and hence,

$$C(v) = \{b\} = \text{conv}\,\{x^{(i)} \mid i \in N\} \qquad \text{whenever } g(N) = 0.$$

It remains to consider the case $g(N) > 0$. Let $x \in C(v)$. Define the real numbers $\alpha_i$, $i \in N$, and the vector $y \in \mathbb{R}^n$ by

$$\alpha_i := [g(N)]^{-1}(b_i - x_i) \qquad \text{and} \qquad y := \sum_{j \in N} \alpha_j x^{(j)}.$$

Now it follows from $x \in C(v)$, Lemma III.1.2(i) and (3.2) that $\alpha_i \geq 0$ for all $i \in N$ as well as $\alpha(N) = 1$. Further, it is straightforward to verify the vector equality $y = x$. We conclude that $x \in \text{conv}\,\{x^{(i)} \mid i \in N\}$. So, (i) holds.
(ii) From Theorem III.5.4 and the convexity of the core, we deduce that $v \in C_1^n$ if and only if $\text{conv}\{x^{(i)} \mid i \in N\} \subset C(v)$. Together with the inverse inclusion obtained in part (i), this explains the statement (ii). □

In Proposition III.5.2 we noticed that the coalition constraints for the core of a 1-convex n-person game can be reduced to the (n-1)-person coalition constraints which is due to the fact that the gap of the grand coalition is at most the gap of any nonempty coalition. Generally speaking, the coalition constraints for the core of an n-person game are superfluous for the coalitions whose gaps are at least the gap of the grand coalition, on the understanding that the (n-1)-person coalition constraints are already met.

PROPOSITION 1.2. Let $v \in G^n$ and $x \in I^*(v)$. The following two statements are equivalent.
(i) $x \in C(v)$.
(ii) $x_i \leq b_i^v$ for all $i \in N$ and
$x(S) \geq v(S)$ for all $S \subset N$ satisfying $g^v(S) < g^v(N)$.

PROOF. The implication (i) ⇒ (ii) is directly derived from (2.8) and Lemma III.1.2(i). In order to prove the converse implication, suppose that (ii) holds. We establish that $x(S) \geq v(S)$ for all $S \subset N$ satisfying $g(S) \geq g(N)$. From both $x \in I^*(v)$ and $x_i \leq b_i$ for all $i \in N$, we obtain

$$\begin{aligned} x(S) &= v(N) - x(N-S) \geq v(N) - b(N-S) \\ &= b(S) - g(N) \qquad \text{for all } S \subset N. \end{aligned} \tag{7.1}$$

Now it follows that $x(S) \geq v(S)$ whenever $g(S) \geq g(N)$. As a result, $x(S) \geq v(S)$ for all $S \subset N$. Hence, $x \in C(v)$. □

Remark that Proposition III.5.2 is a direct consequence of the above proposition. Moreover, from the inequality (7.1) we deduce that the structure of the core is invariant under the replacement of the worth $v(S)$ by the (larger) amount $b^v(S) - g^v(N)$ for any nontrivial coalition S whose gap is at least the gap of the grand coalition N. This invariance of the core gives rise to look at an associated cover that majorizes the original game. For that purpose, we divide the coalitions into two types according to their size with respect to an arbitrary natural number k. Coalitions which have less than k players are equally powerful in the two games. For any coalition S with at least k players, the increase of its power is equal to the nonnegative amount $g^v(S) - g^v(N)$.

DEFINITION 1.3. Let $v \in G^n$ and $k \in \mathbb{N}$. The corresponding game $v_k \in G^n$ is given by

$$\begin{aligned} v_k(S) &:= v(S) && \text{if } |S| < k \\ &= v(N) - \sum_{j\in N-S} b_j^v = \sum_{j\in S} b_j^v - g^v(N) && \text{if } |S| \geq k. \end{aligned}$$

The n-person game $v_k$ is called the <u>k-cover</u> of the game v if the original game v satisfies the condition

$$g^v(S) \geq g^v(N) \qquad \text{for all } S \subset N \text{ with } |S| \geq k. \tag{7.2}$$

Obviously, we have $v_k(N) = v(N)$ and hence, $I^*(v_k) = I^*(v)$. In case (7.2) holds, the term k-cover is explained by the fact that the game $v_k$ majorizes the original game v, i.e.,

$$v_k(S) \geq v(S) \text{ for all } S \subset N \qquad \text{whenever (7.2) holds.}$$

Note that the equality holds in (7.2) for all (n-1)-person coalitions because of Lemma III.1.2(iii) and in addition,

$$v_k(S) = v(S) \text{ for all } S \subset N \qquad \text{whenever } k \geq n-1.$$

In the next lemma we describe the relationships between the upper vectors, the gap functions and the cores respectively of both games. It appears that the involved upper vectors coincide, the gaps of coalitions which have less than k players are not affected, whereas the gaps of coalitions with at least k players in the induced game $v_k$ are equal to the gap of the grand coalition in the game v. Further, the core of the game $v_k$ always contains the core of the game v.

LEMMA 1.4. Let $v \in G^n$ and $k \in \mathbb{N}$. Then

(i) $b^{v_k} = b^v$.

(ii) $g^{v_k}(S) = g^v(S)$ for all $S \subset N$ with $|S| < k$,

$g^{v_k}(S) = g^v(N)$ for all $S \subset N$ with $|S| \geq k$.

(iii) $C(v) \subset C(v_k)$.

(iv) $C(v) = C(v_k)$ whenever (7.2) holds.

PROOF. We always have $v_k(N) = v(N)$ and $v_k(N-\{i\}) = v(N-\{i\})$ for all $i \in N$. Together with (3.1) and (3.2), this implies

$$b_i^{v_k} = v_k(N) - v_k(N-\{i\}) = v(N) - v(N-\{i\}) = b_i^v \quad \text{for } i \in N,$$

$$g^{v_k}(S) = b^{v_k}(S) - v_k(S) = b^v(S) - v(S) = g^v(S) \text{ if } |S| < k,$$

$$g^{v_k}(S) = b^{v_k}(S) - v_k(S) = b^v(S) - v_k(S) = g^v(N) \text{ if } |S| \geq k.$$

So, (i) and (ii) hold. Part (ii) and $g^{v_k}(N) = g^v(N)$ yield

$$g^{v_k}(S) < g^{v_k}(N) \quad \text{iff} \quad |S| < k \text{ and } g^v(S) < g^v(N).$$

By applying Proposition 1.2 to the game $v_k$, it follows that

$$\begin{aligned} x \in C(v_k) \text{ iff } & x(N) = v_k(N),\ x_i \leq b_i^{v_k} \text{ for all } i \in N \text{ and} \\ & x(S) \geq v_k(S) \text{ whenever } |S| < k,\ g^v(S) < g^v(N) \\ \text{iff } & x(N) = v(N),\ x_i \leq b_i^v \text{ for all } i \in N \text{ and} \\ & x(S) \geq v(S) \text{ whenever } |S| < k,\ g^v(S) < g^v(N). \end{aligned}$$

From this and Proposition 1.2 applied to the game $v$, we conclude that $x \in C(v)$ implies $x \in C(v_k)$. In particular, we get the equivalence $x \in C(v)$ iff $x \in C(v_k)$ when the game $v$ satisfies the condition (7.2). So, (iii) and (iv) hold. □

According to Lemma 1.4(iv), a game and its k-cover possess identical cores. The other sections of this chapter are devoted to the study of the convexity of the k-cover of a game which would imply that the core of the (nonconvex) game itself has a very regular structure.

DEFINITION 1.5. Let $k \in \mathbb{N}$. The class $C_k^n$ of <u>k-convex n-person games</u> is given by

$$C_k^n := \{v \in G^n \mid \text{the k-cover } v_k \text{ of the game } v \text{ is convex}\}.$$

For any $k \geq n-1$ we noted $v_k = v$, while the condition (7.2) trivially holds. Thus, for any $k \geq n-1$ we conclude that an n-person game is k-convex if and only if the game itself is convex. Therefore, $C_k^n = C^n$ for all $k \in \mathbb{N}$ with $k \geq n-1$.
In case $k = 1$, then $v_1(S) = b^v(S) - g^v(N)$ for all $S \neq \emptyset$ and hence, the convexity condition (5.2) applied to the game $v_1$ reduces to the simple condition $g^v(N) \geq 0$. From this and the condition (7.2), it follows that the above notion of 1-convexity agrees with the 1-convexity condition (3.11).

EXAMPLE 1.6. Consider again the 3-person game v of (2.10) where $0.5 < \alpha < 1$. The game v itself, the corresponding gap function $g^v$ and the two associated games $v_1$, $v_2$ are listed in Table 7.1. Notice that the upper vector $b^v = (1,1-\alpha,1-\alpha)$ and the gap $g^v(N) = 2-2\alpha$. Clearly, the associated game $v_2$ is equal to the original game v which is not convex because $v(12) - v(1) > v(123) - v(13)$. Hence, the 3-person game v is not k-convex for any $k \geq 2$.

| S | 1 | 2 | 3 | 12 | 13 | 23 | N |
|---|---|---|---|---|---|---|---|
| $v(S)$ | 0 | 0 | 0 | $\alpha$ | $\alpha$ | 0 | 1 |
| $g^v(S)$ | 1 | $1-\alpha$ | $1-\alpha$ | $2-2\alpha$ | $2-2\alpha$ | $2-2\alpha$ | $2-2\alpha$ |
| $v_1(S)$ | $2\alpha-1$ | $\alpha-1$ | $\alpha-1$ | $\alpha$ | $\alpha$ | 0 | 1 |
| $v_2(S)$ | 0 | 0 | 0 | $\alpha$ | $\alpha$ | 0 | 1 |

TABLE 7.1. The 3-person game v of (2.10) where $0.5 < \alpha < 1$, the corresponding gap function $g^v$ as well as the associated games $v_1$ and $v_2$ of Definition 1.3.

The core $C(v_1)$ of the game $v_1$ is drawn in Figure 7.2 where

$$C(v_1) = \text{conv}\,\{(1,\alpha-1,1-\alpha),\ (1,1-\alpha,\alpha-1),\ (2\alpha-1,1-\alpha,1-\alpha)\}.$$

The core $C(v_1)$ is a triangle inside the imputation set

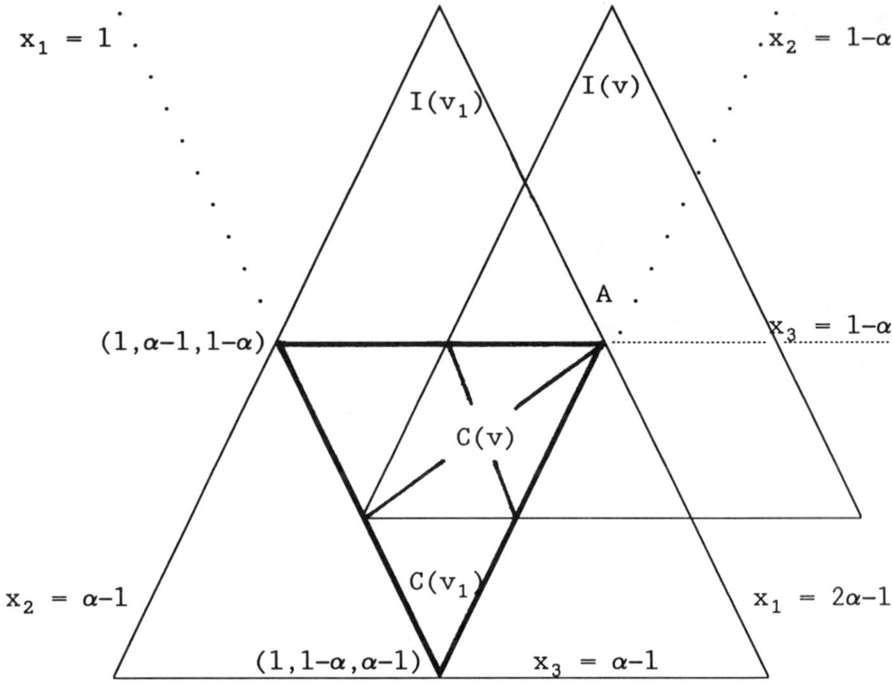


FIGURE 7.2. The geometric position of the core $C(v_1)$ of the associated game $v_1$ with respect to the core $C(v)$ of the game v of (2.10) where $0.5 < \alpha < 1$. The indicated triangle represents the core of the game $v_1$ where the point $A = (2\alpha-1, 1-\alpha, 1-\alpha)$.

$I(v_1)$ such that each vertex of the core is a boundary point of the imputation set. Notice that the imputation sets of both games v and $v_1$ do not coincide because the one-person coalitions have different worths in both games. From the Figures 2.1 and 7.2 we deduce that the core of the original game v is strictly included in the core of the associated game $v_1$, where the core $C(v_1)$ of the game $v_1$ is precisely the convex hull of the set of all three adjusted efficient upper vectors $b^v - g^v(N)e^i$, $i \in N$, in the game v. From this and Theorem 1.1(ii) as well, we conclude that the game v is not 1-convex. Nevertheless, the game $v_1$ is convex by a straightforward verification of the convexity condition (5.2) applied to the game $v_1$. In view of Definition 1.5, the game v fails to be 1-convex since the game v does not satisfy the condition (7.2) applied to k = 1 because of the strict inequality $g^v(N) > g^v(\{2\})$.

EXAMPLE 1.7. Consider again the 1-convex 3-person game v of Example III.5.6. Then the associated game $v_1$ is given by

$$v_1(\{2\}) = v_1(\{3\}) = 6 \quad \text{and} \quad v_1(S) = v(S) \quad \text{otherwise.}$$

The geometric positions of both the imputation set and the core of the game $v_1$ with respect to the imputation set of the game v are drawn in Figure 7.3. Notice that (7.2) holds for all $k \in \mathbb{N}$ and hence, $C(v_k) = C(v)$ for all $k \in \mathbb{N}$ by using Lemma 1.4(iv). From the structure of the core inside the imputation set as shown in the Figures 3.2 and 7.3 respectively, we conclude that the original game v is not convex, whereas the 1-cover $v_1$ is a convex game. So, the 3-person game v is k-convex iff k = 1.

Due to Theorem 1.1, the core of a 1-convex n-person game v degenerates into a singleton if and only if all adjusted efficient upper vectors $b^v - g^v(N)e^i$, $i \in N$, in the game v coincide, i.e., the gap $g^v(N)$ of the grand coalition equals zero. The geometric positions of the vertices of the core of a 1-convex 3-person game with respect to the imputation set are drawn in Figure 7.4.

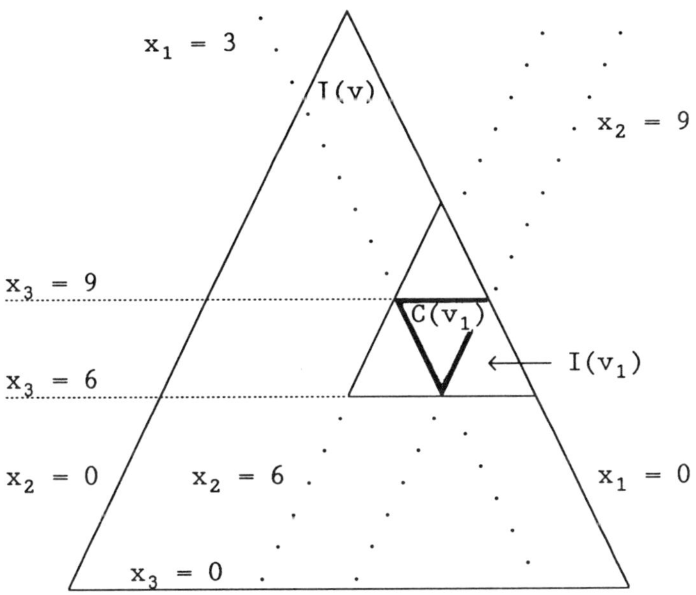


FIGURE 7.3. The indicated triangle represents the core $C(v_1)$ of the associated game $v_1$ inside its imputation set $I(v_1)$ as well as the imputation set I(v) of the 1-convex 3-person game v of Example III.5.6.

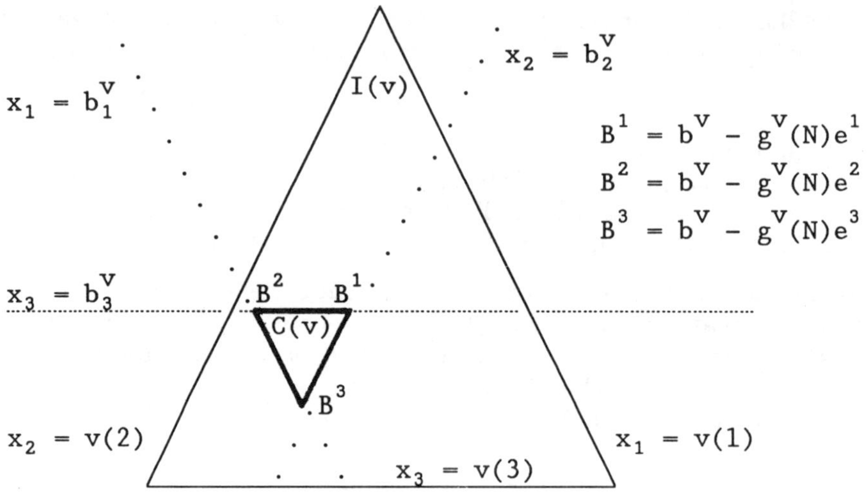


FIGURE 7.4. The geometric positions of the three adjusted efficient upper vectors $b^v - g^v(N)e^i$, $1 \le i \le 3$, in a 1-convex 3-person game v with respect to the imputation set I(v). Usually, each vertex of the core is interior to the imputation set of the 1-convex 3-person game.

## 2. Characterizations of k-convexity in terms of the gap function

In the previous section we introduced the k-convexity of a game as the convexity of its k-cover. In this section we reformulate the k-convexity of a game in terms of the corresponding gap function. We treat two reformulations of k-convexity which can be seen as generalizations of the two equivalent convexity conditions (5.5) and (5.3) respectively in terms of the gap function.

THEOREM 2.1. Let $v \in G^n$ and $k \in \mathbb{N}$. The game v is k-convex iff the game v satisfies the following four conditions:

$$g^v(S) \ge g^v(N) \quad \text{for all } S \subset N \text{ with } |S| \ge k \tag{7.2}$$

$$g^v(N) \ge g^v(S) \quad \text{for all } S \subset N \text{ with } |S| = k-1 \tag{7.3}$$

$$g^v(S) + g^v(T) \ge g^v(S \cup T) + g^v(S \cap T) \tag{7.4}$$
for all S, $T \subset N$ such that $|S \cup T| \le k-1$

$$g^v(S) + g^v(T) \ge g^v(N) + g^v(S \cap T) \tag{7.5}$$
for all S, $T \subset N$ such that $|S \cup T| = k$.

THEOREM 2.2. Let $v \in G^n$ and $k \in \mathbb{N}$. The game v is k-convex iff the game v satisfies the following four conditions:

$$g^v(S) \geq g^v(N) \qquad \text{for all } S \subset N \text{ with } |S| \geq k \tag{7.2}$$

$$g^v(N) \geq g^v(S) \qquad \text{for all } S \subset N \text{ with } |S| = k-1 \tag{7.3}$$

$$g^v(S \cup \{i\}) - g^v(S) \geq g^v(T \cup \{i\}) - g^v(T) \tag{7.6}$$
for all $i \in N$ and all $S, T \subset N$
such that $S \subset T \subset N-\{i\}$ and $|T| \leq k-2$

$$g^v(S \cup \{i\}) - g^v(S) \geq g^v(N) - g^v(T) \tag{7.7}$$
for all $i \in N$ and all $S, T \subset N$
such that $S \subset T \subset N-\{i\}$ and $|T| = k-1$.

PROOF of the Theorems 2.1 and 2.2. Let $v \in G^n$ and $k \in \mathbb{N}$.
(i) Suppose that the game v is k-convex. Then (7.2) holds and its k-cover is convex. We show that (7.3)-(7.5) hold. By (5.5), the convexity property for the k-cover $v_k$ can be written as follows: for all $S, T \subset N$

$$g^{v_k}(S) + g^{v_k}(T) \geq g^{v_k}(S \cup T) + g^{v_k}(S \cap T).$$

From this and Lemma 1.4(ii), it is obvious that (7.4) holds and in addition, (7.5) holds for all $S, T \subset N$ satisfying $|S \cup T| = k$, $|S| \neq k$, and $|T| \neq k$. In case $|S \cup T| = k$ such that $|S| = k$ or $|T| = k$, then (7.5) holds because of (7.2). It remains to prove (7.3). In view of Proposition V.1.3(i), the convexity of the game $v_k$ implies that

$$g^{v_k}(S) \leq g^{v_k}(N) \qquad \text{for all } S \subset N. \text{ Hence,}$$

$$g^v(S) = g^{v_k}(S) \leq g^{v_k}(N) = g^v(N) \qquad \text{whenever } |S| = k-1$$

where the two equalities follow from Lemma 1.4(ii). We conclude that (7.2)-(7.5) hold whenever v is k-convex.
(ii) We show that (7.4) and (7.5) imply (7.6) and (7.7). Suppose that (7.4) and (7.5) hold. Let $i \in N$ and $Q, R \subset N$ be such that $Q \subset R \subset N-\{i\}$. Put $S := R$ and $T := Q \cup \{i\}$. Clearly, $S \cup T = R \cup \{i\}$ and $S \cap T = Q$. In case $|R| \leq k-2$, then $|S \cup T| \leq k-1$ and it follows from (7.4) that

$$g^v(R) + g^v(Q \cup \{i\}) \geq g^v(R \cup \{i\}) + g^v(Q).$$

Thus, (7.6) holds. In case $|R| = k-1$, then $|S \cup T| = k$ and we obtain from (7.5) that

$$g^v(R) + g^v(Q \cup \{i\}) \geq g^v(N) + g^v(Q).$$

Thus, (7.7) holds. So, (7.4)-(7.5) imply (7.6)-(7.7).

(iii) Suppose that (7.2), (7.3), (7.6) and (7.7) hold. We show that the k-cover $v_k$ is convex. We verify the convexity condition (5.3) for the game $v_k$, i.e.,

$$g^{v_k}(S \cup \{i\}) - g^{v_k}(S) \geq g^{v_k}(T \cup \{i\}) - g^{v_k}(T) \qquad (7.8)$$

for all $i \in N$ and all $S, T \subset N$ such that $S \subset T \subset N-\{i\}$. For that purpose, we make use of Lemma 1.4(ii). Let $i \in N$ and $S \subset T \subset N-\{i\}$ where $S \neq T$. In case $|T| \leq k-2$, then (7.8) reduces to (7.6). In case $|T| = k-1$, then (7.8) reduces to (7.7). In case $|T| \geq k$, then the condition (7.8) reduces to

$$g^{v_k}(S \cup \{i\}) \geq g^{v_k}(S)$$

which inequality is an equality whenever $|S| \geq k$, while the involved inequality is due to (7.3) whenever $|S| = k-1$. So, it remains to prove that

$$g^{v}(S \cup \{i\}) \geq g^{v}(S) \qquad \text{whenever } |S| \leq k-2. \qquad (7.9)$$

In case $|S| \leq k-2$, then there exists a coalition $R \subset N$ such that $S \subset R \subset N-\{i\}$ and $|R| = k-1$. Now it follows from (7.7) and (7.3) respectively that

$$g^{v}(S \cup \{i\}) - g^{v}(S) \geq g^{v}(N) - g^{v}(R) \geq 0.$$

Thus, (7.9) holds and hence, (7.8) holds. We conclude that v is k-convex whenever (7.2), (7.3), (7.6) and (7.7) hold.
(iv) The equivalences mentioned in the two theorems follow immediately from the parts (i)-(iii). □

We usually prefer the two k-convexity conditions (7.6) and (7.7) to the pair of equivalent conditions (7.4) and (7.5). An interpretation of the conditions (7.6) and (7.7) is as follows. Suppose that the formation of the grand coalition N is seen as the sequential process where players enter one by one until a group of k-1 players is formed and subsequently, all other players enter together. According to this process, single players merely join coalitions with at most k-2 members. In view of this specific formation of the grand coalition, the k-convexity conditions (7.6) and (7.7) express that the gap function satisfies nonincreasing marginal returns with respect to the coalition size.

By Proposition V.1.3(i), the gap function of a convex game is monotonic. For any k-convex n-person game where $1 \leq k \leq n-2$, the gaps of the coalitions are divided into three levels. Here the lower level consists of gaps of coalitions with at most k-1 players and this lower level possesses the monotonicity property with respect to the coalition size (see (7.9)). The middle level consists of gaps of the grand coalition N as well as the (n-1)-person coalitions. The upper and final level consists of gaps of the remaining coalitions with coalition size at least k and at most n-2. The middle level only contains equal gaps, but at the upper level the gaps are arbitrary. It follows that the gap function of a k-convex game is nonnegative since the gap of the empty coalition is minimal among all gaps.

From the above division of the gaps into three levels, we also deduce that k-convexity and m-convexity where $k \neq m$ are in general contradictory notions.

EXAMPLE 2.3. Consider again the 8-person game v of Example II.6.11 where $\alpha$ is any nonnegative real number. Then we get that $b_i = 6$ for all $i \in N$ and

$$
\begin{aligned}
&g(\{i\}) = 6 && \text{for all } i \in N, \\
&g(N) = g(N-\{i\}) = 8 && \text{for all } i \in N, \\
&g\big((P \cup \{q\})-\{p\}\big) = 24-\alpha && \text{for all } p \in P \text{ and all } q \in Q, \\
&g\big((Q \cup \{p\})-\{q\}\big) = 24-\alpha && \text{for all } p \in P \text{ and all } q \in Q, \\
&g(S) = 6|S| \geq 12 && \text{for all other } S \neq \emptyset.
\end{aligned}
$$

Therefore, $g(\{i,j\}) > g(N) > g(\{i\})$ for all $i,j \in N$, $i \neq j$. From these strict inequalities and the specific division of the gaps into three levels due to k-convexity, it follows that the game v fails to be k-convex for all $k \neq 2$. In view of Theorem 2.2, for the 2-convexity of the game v it is necessary and sufficient that the following conditions hold:

$$g(S) \geq g(N) \quad \text{and} \quad g(\{i\}) + g(\{j\}) \geq g(N) \geq g(\{i\})$$

$$\text{for all } S \subset N \text{ with } |S| \geq 2 \text{ and for all } i,j \in N,\ i \neq j.$$

From this we conclude that the 8-person game v is 2-convex iff $\alpha \leq 16$.

Finally, we characterize convexity of k-convex games by the fact that the inequalities in (7.2) are equalities. In other words, a k-convex game is also convex if and only if the upper level of gaps coincides with the middle level.

PROPOSITION 2.4. Let $v \in G^n$ and $k \in \mathbb{N}$. The following three statements are equivalent.

(i) v is k-convex and convex.

(ii) v is k-convex and $g^v(S) = g^v(N)$ whenever $|S| \geq k$.

(iii) v is convex and $g^v(S) = g^v(N)$ whenever $|S| \geq k$.

PROOF. We prove (i) $\Rightarrow$ (ii) $\Rightarrow$ (iii) $\Rightarrow$ (i).
(a) Suppose that (i) holds. In view of Proposition V.1.3, the convexity of the game v implies that $g^v(S) \leq g^v(N)$ for all $S \subset N$. From this and the inequality in (7.2), we derive the equality $g^v(S) = g^v(N)$ for all $S \subset N$ such that $|S| \geq k$. So, (i) implies (ii).
(b) Suppose that (ii) holds. Then the k-cover $v_k$ is convex and we have that for all $S \subset N$ with $|S| \geq k$

$$v_k(S) = b^v(S) - g^v(N) = b^v(S) - g^v(S) = v(S).$$

We obtain that $v_k(S) = v(S)$ for all $S \subset N$. Thus, $v_k = v$ and hence, the game v itself is convex. So, (ii) implies (iii).
(c) Suppose that (iii) holds. Then we have $v_k = v$ as shown in part (b). Hence, both the game v and the k-cover $v_k$ are convex games. So, (iii) implies (i). □

COROLLARY 2.5. Let $v \in G^n$ and $k \in \mathbb{N}$ be such that (7.2) holds. Then v is k-convex iff its k-cover $v_k$ is k-convex.

PROOF. By Lemma 1.4(ii), we have that $g^{v_k}(S) = g^{v_k}(N)$ for all $S \subset N$ with $|S| \geq k$. Due to Proposition 2.4(ii)-(iii) applied to the k-cover $v_k$, it follows that k-convexity and convexity are equivalent notions for the k-cover. From this we conclude that the k-convexity of the original game v is equivalent to the k-convexity of its k-cover $v_k$. □

According to the above proposition, a convex game is also k-convex if and only if the corresponding gap function is constant for the coalitions with at least k players. This characterization is used to investigate the k-convexity of the submodular cost function of the airport cost game.

EXAMPLE 2.6. We consider again the airport cost game (N;c) as described in Section I.5. The submodularity of the cost function c and the induced convexity of the corresponding savings game (N;v) are established in Example V.2.4. From both Lemma IV.2.3 and the equalities in (4.13)-(4.14), it follows immediately that the gap function of the savings game v is given by

$$g^v(S) = c(S) \text{ for all } S \subset N \qquad \text{in case } |N_m| \geq 2,$$

$$g^v(S) = c(S) \text{ for all } S \subset N-\{\hat{\imath}\} \qquad \text{in case } N_m = \{\hat{\imath}\}$$

$$= C_{m-1} \text{ for all } S \subset N \text{ with } \hat{\imath} \in S.$$

For the nontrivial situation with at least two landings at the airport, we distinguish three cases in order to study the gap function. For each case, the result concerning the k-convexity of the convex savings game v is derived from Proposition 2.4(i) and (iii). The results are listed in Table 7.5. On the understanding that a unique largest plane of type m is regarded as a plane of type m-1, we give an interpretation of the result for the situation with at least two different types of aircraft. For the k-convexity of the savings game, it is necessary and sufficient that the natural number k exceeds the number of landings by planes of all but the largest type. In particular, the k-convexity does not depend upon the costs $C_j$, $1 \leq j \leq m$, where each cost $C_j$ represents the cost of a runway adequate for planes of type j.

| case | $\lvert N_m\rvert \geq 2$, m=1 or $\lvert N_m\rvert = 1$, m=2 | $\lvert N_m\rvert \geq 2$ and $m \geq 2$ | $\lvert N_m\rvert = 1$ and $m \geq 3$ |
|---|---|---|---|
| gap function | $g^v(S) = C_1$ for all $S \neq \emptyset$ | $g^v(\bigcup_{j=1}^{m-1} N_j) = C_{m-1}$ <br> $g^v(S) = C_m$ if $\lvert S\rvert > \lvert N\rvert - \lvert N_m\rvert$ | $g^v(\bigcup_{j=1}^{m-2} N_j) = C_{m-2}$ <br> $g^v(S) = C_{m-1}$ if $\lvert S\rvert > \lvert N\rvert - \lvert N_{m-1}\rvert - 1$ |
| k-convexity | v is k-convex for all $k \in \mathbb{N}$ | v is k-convex iff $k > \lvert N\rvert - \lvert N_m\rvert$ | v is k-convex iff $k > \lvert N\rvert - \lvert N_{m-1}\rvert - 1$ |

TABLE 7.5. The results concerning the k-convexity of the savings game (N;v) associated with the airport cost game (N;c).

### 3. Examples of k-convex games

We investigate the k-convexity of simple games as well as symmetric games. Concerning the class of simple games, a game with a unique veto player is k-convex either for all $k \in \mathbb{N}$ or merely for $k = 1$. Furthermore, a simple n-person game with at least two veto players is k-convex either for all $k \geq n-1$ or for no k at all. As a matter of fact, in case $k \geq 2$, then the k-convexity of a simple n-person game is precisely the convexity of the simple game which is characterized by the fact that the set of all veto players is a powerful coalition in the game.

THEOREM 3.1. Let $v \in S^n$ and $k \in \mathbb{N}$.

(i) If $J^v = \emptyset$, then v is not k-convex.

(ii) If $|J^v| = 1$, then v is 1-convex.

(iii) Let $|J^v| = 1$ and $k \geq 2$.
Then v is k-convex iff v is convex.

(iv) Let $|J^v| \geq 2$.
Then v is k-convex iff v is convex and $k \geq n-1$.

PROOF. From both $v \in S^n$, (3.1) and (3.2) we derive that

$$\begin{aligned} b_i^v &= 1 \qquad \text{if } i \in J^v \\ &= 0 \qquad \text{if } i \in N-J^v, \end{aligned}$$

$$g^v(N) = |J^v| - 1,$$

$$g^v(S) = |S \cap J^v| - v(S) \qquad \text{for all } S \subset N.$$

(i) Suppose $J^v = \emptyset$. Evidently, we obtain $g^v(N) = -1 < 0$. Since the corresponding gap function $g^v$ is not nonnegative, the game v is not k-convex. So, (i) holds.
(ii)-(iii) Suppose $|J^v| = 1$, say $J^v = \{j\}$. Then we have $v(N-\{j\}) = 0$ and therefore, $v(S) = 0$ for all $S \subset N-\{j\}$ by the monotonicity of the simple game v. Now it follows that

$$\begin{aligned} g^v(S) &= 1 - v(S) \qquad \text{if } j \in S \\ &= 0 \qquad \text{if } j \in N-S. \end{aligned}$$

In particular, we get that $g^v(S) \geq 0 = g^v(N)$ for all $S \subset N$ and as such, the game v satisfies the condition (3.11) as well as (7.2) applied to all $k \in \mathbb{N}$. By (3.11), the game v is 1-convex and so, (ii) holds. In order to prove (iii), suppose $k \geq 2$ too. The k-cover $v_k$ is given by

$$\begin{aligned} v_k(S) &= v(S) && \text{if } |S| < k \\ &= 1 && \text{if } |S| \geq k,\ j \in S \\ &= 0 && \text{if } |S| \geq k,\ j \in N-S. \end{aligned}$$

It is straightforward to verify that the game $v_k$ is also a simple game. Then $v_k(J^v) = v(J^v)$ because $|J^v| = 1 < k$ and further, $J^{v_k} = J^v$ as a direct consequence of Lemma 1.4(i). From these two equalities and Theorem V.2.5 applied to the two simple games $v_k$ and $v$ respectively, we conclude that

$$v \text{ is } k\text{-convex} \Leftrightarrow v_k \text{ is convex} \Leftrightarrow v_k(J^{v_k}) = 1$$

$$\Leftrightarrow v_k(J^v) = 1 \Leftrightarrow v(J^v) = 1 \quad \Leftrightarrow v \text{ is convex.}$$

So, (iii) holds as well.
(iv) Let $|J^v| \geq 2$, say $\{i,j\} \subset J^v$ where $i \neq j$. Then we get $v(N-\{i,j\}) = 0$ by the monotonicity of $v$. It follows that

$$g^v(N-\{i,j\}) = |J^v| - 2 < |J^v| - 1 = g^v(N)$$

which strict inequality is in contradiction with (7.2) when $1 \leq k \leq n-2$. Further, for any $k \geq n-1$, the k-convexity of $v$ is equivalent to the convexity of $v$. So, (iv) holds. □

The set of all veto players in the simple n-person game $v$ of (1.4) is a singleton because the chairman is the unique veto player, i.e., $J^v = \{1\}$. Note that $v(J^v) = 1$ iff $m = 1$. The convexity condition $m = 1$ expresses that the vote of the chairman is necessary and sufficient to pass a bill. In case $m > 1$, then the simple game $v$ is not convex and hence, the game $v$ is k-convex merely for $k = 1$. In case $m = 1$, then the game $v$ is k-convex for all $k \in \mathbb{N}$.

EXAMPLE 3.2. We consider a production economy consisting of traders of the same type where each trader initially holds one unit of raw material. Any group of traders can produce an output by using their own material on the understanding that one unit of the raw material is needed to produce one unit of a certain product. Moreover, it is supposed that a group which is not too small but does not have the majority of all n traders as its members (i.e., its group size is at least m but less than $(n+1)/2$) can not produce more than a group of m-1 traders. One unit of the product can be sold at a net profit of one unit of money.

This economic situation can be modelled as a cooperative game (N;v) where its player set $N = \{1,2,\ldots,n\}$ consists of the n traders and its characteristic function v agrees with the net profit function. That is, the game v is given by

$$v(S) := |S| \qquad \text{if } |S| < m \text{ or } |S| \geq \tfrac{1}{2}(n+1)$$
$$= m-1 \qquad \text{if } m \leq |S| < \tfrac{1}{2}(n+1). \tag{7.10}$$

The game v of (7.10) is called symmetric since the worth of any coalition depends upon the coalition size instead of the players in the coalition. Formally, a game $v \in G^n$ is said to be <u>symmetric</u> if

there exists a function $f\colon \{0,1,\ldots,n\} \to \mathbb{R}$

such that $\quad v(S) = f(|S|) \quad$ for all $S \subset N$.

In the remainder of the section we characterize k-convexity of symmetric games. For that purpose, we first reformulate the k-convexity of any game in terms of the characteristic function itself as well as the upper vector of the game.

THEOREM 3.3. Let $v \in G^n$ and $k \in \mathbb{N}$. The game v is k-convex iff the game v satisfies the following four conditions:

$$v(N) - v(S) \geq b^v(N-S) \tag{7.11}$$
for all $S \subset N$ with $|S| \geq k$

$$v(N) - v(S) \leq b^v(N-S) \tag{7.12}$$
for all $S \subset N$ with $|S| = k-1$

$$v(S \cup \{i\}) - v(S) \leq v(T \cup \{i\}) - v(T) \tag{7.13}$$
for all $i \in N$ and all $S,\ T \subset N$ such that $S \subset T \subset N-\{i\}$ and $|T| \leq k-2$

$$v(N) - v(S) \geq b^v\big((N-S)-\{i\}\big) + \max_{j \in S}\big[v((S \cup \{i\})-\{j\}) - v(S-\{j\})\big] \tag{7.14}$$
for all $i \in N$ and all $S \subset N$ such that $S \subset N-\{i\}$, $S \neq \emptyset$, and $|S| = k-1$.

PROOF. By (3.2), it is obvious that (7.2), (7.3) and (7.6) respectively are reformulations of (7.11),(7.12) and (7.13) in terms of the corresponding gap function.
Now we prove the equivalence of (7.7) and (7.14) on the understanding that (7.11) and (7.13) are supposed to hold. Note that the relevant condition (7.14) is equivalent to

$$g(N) - g(S) \le \min_{j \in S}\left[g((S \cup \{i\})-\{j\}) - g(S-\{j\})\right] \quad (7.15)$$

for all $i \in N$ and all $S \subset N$
such that $S \subset N-\{i\}$, $S \neq \emptyset$, and $|S| = k-1$.

(i) Suppose that (7.7) holds. Let $i \in N$, $S \subset N-\{i\}$, $S \neq \emptyset$, $|S| = k-1$, and $j \in S$. Then it follows from (7.7) that

$$g(N) - g(S) \le g((S \cup \{i\})-\{j\}) - g(S-\{j\}).$$

Thus, (7.15) holds. So, (7.7) implies (7.14).
(ii) Suppose that (7.11), (7.13) and (7.14) hold. Clearly, the equivalent conditions (7.2), (7.6) and (7.15) in terms of the corresponding gap function are valid. We show (7.7). Let $i \in N$, $S \subset T \subset N-\{i\}$, and $|T| = k-1$. In case $S = T$, then (7.7) reduces to $g(T \cup \{i\}) \ge g(N)$ which inequality holds because of (7.2). In case $S \neq T$, then there exists $j \in T-S$ and it follows from (7.15) and (7.6) respectively that

$$g(N) - g(T) \le g((T \cup \{i\})-\{j\}) - g(T-\{j\})$$

$$\le g(S \cup \{i\}) - g(S).$$

Thus, (7.7) holds. So, the three conditions (7.11), (7.13) and (7.14) imply (7.7).
(iii) In the parts (i)-(ii) we have shown the equivalence of (7.7) and (7.14) on the understanding that both (7.11) and (7.13) are supposed to hold. Together with Theorem 2.2, this completes the proof. □

An interpretation of the resulting k-convexity conditions (7.11)-(7.14) is as follows. The condition (7.13) requires the convexity of the game v with respect to the coalition size up to k. According to (7.11), the total savings v(N) of the grand coalition N can be distributed among the players in such a way that any coalition S with at least k players receives at least its worth and any player outside the coalition S gets at least his marginal contribution to the grand coalition N. This specific way to distribute the amount v(N) among the players is not possible (in the weak sense) any more for coalitions with precisely k-1 players because of (7.12). However, the condition (7.14) expresses that the specific distribution of the amount v(N) is still possible for any coalition S with k-1 players whenever the marginal contribution to the coalition N of an arbitrary player outside the coalition S is replaced by his marginal contribution with respect to an arbitrary coalition which is obtained from S by removing one member.

The k-convexity conditions (7.11)-(7.14) for a symmetric n-person game can also be reformulated in terms of the function associated with the symmetric game. The resulting k-convexity conditions for the associated function follow immediately from Theorem 3.3 and are mentioned in the next proposition.

PROPOSITION 3.4. Let $v \in G^n$, $k \in \mathbb{N}$ and $f: \{0,1,\ldots,n\} \to \mathbb{R}$ be such that $1 \le k \le n$ and

$$v(S) = f(|S|) \text{ for all } S \subset N. \quad \text{Put } \Delta_n(f) := f(n) - f(n-1).$$

Then the symmetric game v is k-convex iff the function f satisfies the following four conditions:

$$f(n) - f(s) \ge (n-s)\Delta_n(f) \quad \text{for all } k \le s \le n \tag{7.16}$$

$$f(n) - f(k-1) \le (n-k+1)\Delta_n(f) \tag{7.17}$$

$$f(s+1) - f(s) \le f(t+1) - f(t) \quad \text{for all } 0 \le s \le t \le k-2 \tag{7.18}$$

$$f(n) - f(k-1) \ge (n-k)\Delta_n(f) + f(k-1) - f(k-2) \quad \text{in case } k > 1. \tag{7.19}$$

A function f that satisfies the conditions (7.16)-(7.19) is said to be <u>k-convex</u>. Clearly, the notion of n-convexity for a function $f: \{0,1,\ldots,n\} \to \mathbb{R}$ agrees with the convexity for the function f as described by the symmetric version of the convexity condition (5.2), i.e.,

$$f(s+1) - f(s) \le f(t+1) - f(t) \quad \text{for all } 0 \le s \le t \le n-1.$$

The convexity of a function f expresses that the marginal returns $f(t+1) - f(t)$, $0 \le t \le n-1$, of the function form a nondecreasing sequence. The k-convexity condition (7.18) for a function $f: \{0,1,\ldots,n\} \to \mathbb{R}$ requires the convexity of the restriction of the function to $\{0,1,\ldots,k-1\}$. The other three k-convexity conditions can be illustrated by means of the geometric positions of certain points with respect to the straight line $\ell_f$ through the points $(n, f(n))$ as well as $(n-1, f(n-1))$. So, the line $\ell_f: \mathbb{R} \to \mathbb{R}$ is given by

$$\ell_f(x) := f(n) + \Delta_n(f)(x - n) \quad \text{for all } x \in \mathbb{R}.$$

The reformulations of the conditions (7.16), (7.17) and (7.19) respectively in terms of the line $\ell_f$ are as follows:

$$f(s) \le \ell_f(s) \quad \text{for all } k \le s \le n \tag{7.20}$$

$$f(k-1) \ge \ell_f(k-1) \tag{7.21}$$

$$2f(k-1) - f(k-2) \le \ell_f(k) \quad \text{in case } k > 1. \tag{7.22}$$

The condition (7.20) expresses that all points $(s, f(s))$ where $k \le s \le n$ lie below the associated line, whereas the point $(k-1, f(k-1))$ lies above the line because of (7.21). Moreover, the condition (7.22) requires that for any $k \ne 1$, the point $(k, 2f(k-1) - f(k-2))$ lies below the line. As a matter of fact, the number k is a critical number for a k-convex function f on $\{0,1,\dots,n\}$ since the function lies above the associated line on $\{0,1,\dots,k-1\}$, but below this line on $\{k,k+1,\dots,n\}$.

LEMMA 3.5. Let $k \in \mathbb{N}$ and $f: \{0,1,\dots,n\} \to \mathbb{R}$ be k-convex such that $1 \le k \le n$. Then

$$f(s) \ge \ell_f(s) \quad \text{for all } 0 \le s \le k-1,$$

$$f(s) \le \ell_f(s) \quad \text{for all } k \le s \le n.$$

PROOF. By the k-convexity of f, we have that (7.20) and (7.21) hold. So, it remains to prove that $\ell_f(s) \le f(s)$ for all $0 \le s < k-1$. Let $0 \le s < k-1$. Then we obtain $k > 1$ and

$$f(k-1) - f(k-2) \le f(n) - f(k-1) - (n-k)\Delta_n(f) \le \Delta_n(f)$$

where the two inequalities follow from (7.19) and (7.17) respectively. Further, from the convexity condition (7.18) we deduce

$$f(k-1) - f(s) = \sum_{t=s}^{k-2} \left[f(t+1) - f(t)\right]$$

$$\le \sum_{t=s}^{k-2} \left[f(k-1) - f(k-2)\right] = (k-1-s)\left[f(k-1) - f(k-2)\right].$$

Now it follows that $f(k-1) - f(s) \le (k-1-s)\Delta_n(f)$. Together with the condition (7.17), this inequality yields

$$f(n) - f(s) = f(n) - f(k-1) + f(k-1) - f(s)$$

$$\le (n-k+1)\Delta_n(f) + (k-1-s)\Delta_n(f) = (n-s)\Delta_n(f).$$ Hence,

$$\ell_f(s) = f(n) + (s-n)\Delta_n(f) \le f(s).$$ □

LEMMA 3.6. Let $k \in \mathbb{N}$ and $f: \{0,1,\ldots,n\} \to \mathbb{R}$ be k-convex such that $f(0) = 0$ and $1 \le k \le n$. Then

$$s^{-1}f(s) \le n^{-1}f(n) \qquad \text{for all } 1 \le s \le n.$$

PROOF. By Lemma 3.5, we have $\ell_f(0) \le f(0)$ which inequality is equivalent to $f(n) \le n\Delta_n(f)$. This implies that

$$\ell_f(s) = f(n) + (s-n)\Delta_n(f) \le n^{-1}sf(n) \text{ for all } 0 \le s \le n.$$

By using again Lemma 3.5, we obtain that

$$f(s) \le \ell_f(s) \le n^{-1}sf(n) \qquad \text{for all } k \le s \le n. \tag{7.23}$$

If $k = 1$, then the lemma is a direct consequence of (7.23). So, it remains to consider the case $k > 1$.

(i) From the convexity condition (7.18) we deduce that for all $1 \le s \le k-2$

$f(s) = f(s) - f(0) \le s\big(f(s+1) - f(s)\big)$ or equivalently,

$s^{-1}f(s) \le (s+1)^{-1}f(s+1)$. Particularly, we get that

$s^{-1}f(s) \le (k-1)^{-1}f(k-1)$ for all $1 \le s \le k-1$.

(ii) We have $\ell_f(k) \le n^{-1}kf(n)$ and further, part (i) yields

$$f(k-2) \le (k-1)^{-1}(k-2)f(k-1).$$

Together with (7.22), the two obtained inequalities imply

$$2f(k-1) \le \ell_f(k) + f(k-2)$$
$$\le n^{-1}kf(n) + (k-1)^{-1}(k-2)f(k-1) \qquad \text{or equivalently,}$$
$$(k-1)^{-1}f(k-1) \le n^{-1}f(n).$$

(iii) From the parts (i)-(ii) we conclude that

$$s^{-1}f(s) \le (k-1)^{-1}f(k-1) \le n^{-1}f(n) \quad \text{for all } 1 \le s \le k-1.$$

Together with (7.23), this completes the proof. □

Lemma 3.6 states that a k-convex function f with $f(0) = 0$ lies completely below the straight line through the points $(0, 0)$ and $(n, f(n))$.

EXAMPLE 3.7. Let the symmetric 4-person game v be given by

$$v(S) = 2|S| - 3 \qquad \text{if } |S| \ge 3$$
$$= 0 \qquad \text{otherwise.}$$

The associated function $f: \{0,1,2,3,4\} \to \mathbb{R}$ as well as the straight line $\ell_f: \mathbb{R} \to \mathbb{R}$ are given by

$$f(0) = f(1) = f(2) = 0, \quad f(3) = 3, \quad f(4) = 5,$$

$$\ell_f(x) := 2x - 3 \qquad \text{for all } x \in \mathbb{R}.$$

Both the function f and the line $\ell_f$ are drawn in Figure 7.6 from which we deduce that the point $(1, f(1))$ lies strictly above the line, whereas the point $(2, f(2))$ lies strictly below the line. So, the number $k = 2$ is a critical number and we conclude from Lemma 3.5 that the function f fails to be k-convex when $k \neq 2$. It is left to the reader to verify the 2-convexity conditions (7.20)-(7.22) for the function f. Thus, the symmetric game v is k-convex merely for $k = 2$.

EXAMPLE 3.8. Consider again the game v of Example 3.2 that arises from a production economy based on majority voting. In view of (7.10), the n-person game v is symmetric and the associated function $f: \{0,1,\ldots,n\} \to \mathbb{R}$ is given by

$$f(s) := s \qquad \text{if } s < m \text{ or } s \geq \tfrac{1}{2}(n+1)$$
$$= m-1 \qquad \text{if } m \leq s < \tfrac{1}{2}(n+1).$$

Here the numbers $m, n \in \mathbb{N}$ are such that $n \geq 3$ and $n > 2m-1$. In particular, we have $f(n) = n$, $f(n-1) = n-1$ and hence,

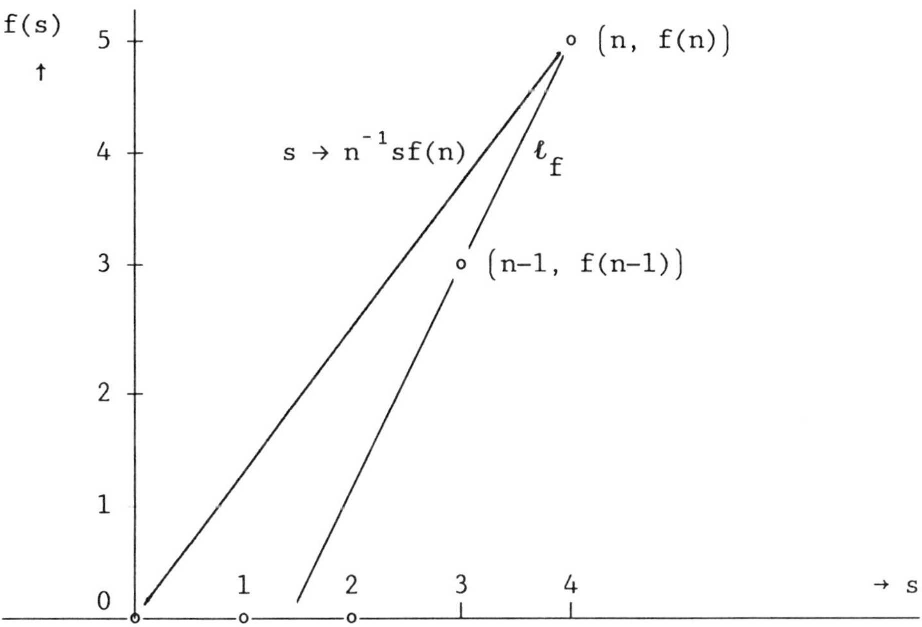


FIGURE 7.6. The function $f: \{0,1,2,3,4\} \to \mathbb{R}$ as well as the straight line $\ell_f: \mathbb{R} \to \mathbb{R}$ corresponding to the symmetric 4-person game v of Example 3.7.

the straight line $\ell_f: \mathbb{R} \to \mathbb{R}$ is given by

$$\ell_f(x) := x \qquad \text{for all } x \in \mathbb{R}.$$

Both the function f and the line $\ell_f$ are drawn in Figure 7.7 from which we deduce that the function f as a whole lies below the line $\ell_f$, i.e., $f(s) \le \ell_f(s)$ for all $0 \le s \le n$. Especially, the point $(m, f(m))$ lies strictly below the associated line. From this fact and Lemma 3.5, it follows that the requirement $k \le m \le n$ is a necessary condition for the k-convexity of the function f. In case $1 \le k \le m$, then we derive from Figure 7.7 that the restrictions of the function f as well as the associated line to $\{0,1,\ldots,k-1\}$ coincide and so, the three k-convexity conditions (7.18), (7.21) and (7.22) are trivially satisfied by the function f. We conclude that f is a k-convex function iff $1 \le k \le m$. In other words, the symmetric game v of (7.10) is k-convex if and only if the number k does not exceed the number m.

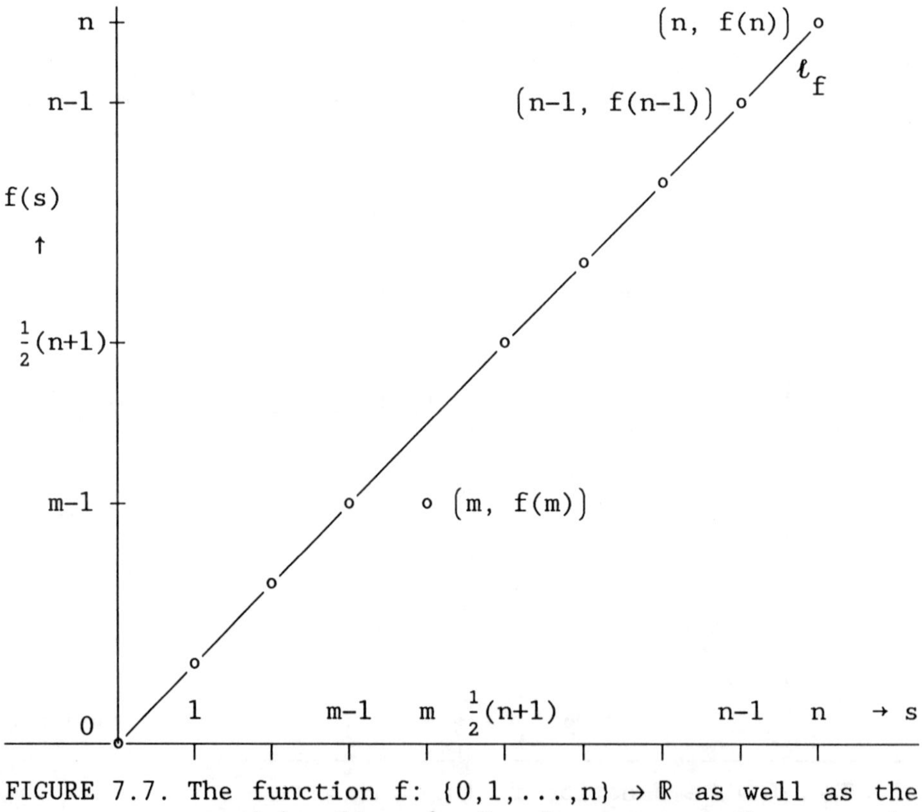


FIGURE 7.7. The function $f: \{0,1,\ldots,n\} \to \mathbb{R}$ as well as the straight line $\ell_f: \mathbb{R} \to \mathbb{R}$ corresponding to the symmetric n-person game v of Example 3.8.

Finally, we study the balancedness of symmetric games. In general, a symmetric game may possess an empty core. It appears that the nonemptiness of the core of a symmetric n-person game v can be characterized by the fact that the symmetric payoff vector $n^{-1}v(N)1_n$ belongs to the core of the game v. A simple reformulation of this characterization in terms of the function associated with the symmetric game is stated in the next theorem.

THEOREM 3.9. Let $v \in G^n$ and $f: \{0,1,\dots,n\} \to \mathbb{R}$ be such that $v(S) = f(|S|)$ for all $S \subset N$. Then

$$C(v) \neq \emptyset \quad \text{iff} \quad s^{-1}f(s) \le n^{-1}f(n) \text{ for all } 1 \le s \le n.$$

PROOF. (i) Suppose that $s^{-1}f(s) \le n^{-1}f(n)$ for all $1 \le s \le n$. Define the vector $x \in \mathbb{R}^n$ by $x_i := n^{-1}f(n)$ for all $i \in N$. Now it follows that $x(N) = f(n) = v(N)$ as well as

$$x(S) = n^{-1}|S|f(n) \ge f(|S|) = v(S) \qquad \text{for all } S \subset N.$$

So, $x \in C(v)$ by (2.8). Hence, $C(v) \neq \emptyset$.
(ii) To prove the converse statement, suppose $C(v) \neq \emptyset$. Let $1 \le s \le n$. We show $s^{-1}f(s) \le n^{-1}f(n)$.
The set of all coalitions of size s is denoted by $W_s$, i.e.,

$$W_s := \{S \mid S \subset N, |S| = s\}. \qquad \text{Clearly, } |W_s| = \binom{n}{s}.$$

Choose $y \in C(v)$. Then $y(S) \ge v(S) = f(s)$ for all $S \in W_s$. Summing up the inequalities over all coalitions of size s, we obtain

$$\binom{n}{s}f(s) = \sum_{S\in W_s} f(s) \le \sum_{S\in W_s} y(S) = \sum_{S\in W_s}\sum_{j\in N} y_j 1_S(j)$$

$$= \sum_{j\in N} y_j \sum_{S\in W_s} 1_S(j) = \sum_{j\in N} y_j \cdot \binom{n-1}{s-1} = \binom{n-1}{s-1} y(N).$$

Together with the equality $y(N) = v(N) = f(n)$, this yields

$$\binom{n}{s}f(s) \le \binom{n-1}{s-1}f(n) \qquad \text{or equivalently,}$$

$$s^{-1}f(s) \le n^{-1}f(n). \qquad \text{This completes the proof. } \square$$

From Theorem 3.9 and Lemma 3.6, we directly conclude that a symmetric k-convex game possesses a nonempty core. The next section is devoted to a study of the structure of the core of k-convex games.

### 4. The core of a k-convex game

For the convexity of an n-person game it is necessary and sufficient that the core of the given game contains all n! marginal worth vectors and similarly, the 1-convexity of an n-person game is characterized by the easy fact that all n adjusted efficient upper vectors belong to the core of the game. In this section we characterize the k-convexity of a game in a similar way by means of the structure of the core of the game. For that purpose, we introduce a notion which is composed of the ideas of the upper vector and a marginal worth vector.
The notion in question prescribes a specific distribution of the total savings v(N) of the grand coalition among the players. The division of the amount v(N) among the players is based on the formation of the grand coalition N as the sequential process where players enter one by one until a group of k-1 players is formed and subsequently, all other players enter together. The order of entrance of players is described with the aid of an ordering $\theta$ of the player set. The single players i who enter one by one (i.e., $\theta(i) < k$) receive their marginal contributions $v(P_i^\theta \cup \{i\}) - v(P_i^\theta)$ with respect to the coalition of their predecessors. All other players with the exception of their leader j (i.e., $\theta(j) = k$) get their marginal contributions with respect to the grand coalition or equivalently, these players are paid according to the upper vector $b^v$. The leader receives the remaining part of the total amount v(N).
Formally, the above distribution of the total savings among the players in a game is defined as follows.

DEFINITION 4.1. Let $v \in G^n$, $k \in \mathbb{N}$ and $\theta \in \Theta^n$.
The <u>adjusted marginal worth vector</u> $x^{\theta,k}(v) \in \mathbb{R}^n$ with respect to the natural number k and the ordering $\theta$ in the game v is given by

$$\begin{aligned} x_i^{\theta,k}(v) &:= v(P_i^\theta \cup \{i\}) - v(P_i^\theta) && \text{if } \theta(i) < k \\ &= b_i^v && \text{if } \theta(i) > k \\ &= v(N) - v(P_i^\theta) - \sum_{j;\theta(j)>k} b_j^v && \text{if } \theta(i) = k. \end{aligned} \tag{7.24}$$

In case $k \geq n-1$ and $k = 1$ respectively, then the adjusted marginal worth vectors in a game are precisely the original marginal worth vectors and the adjusted efficient upper vectors in the game itself. In fact, each adjusted marginal worth vector in a game corresponds to an original marginal worth vector in the associated game $v_k$ of Definition 1.3.

LEMMA 4.2. Let $v \in G^n$, $k \in \mathbb{N}$ and $\theta \in \Theta^n$. Then

(i) $x^{\theta,k}(v) = x^{\theta}(v_k)$.

(ii) $x^{\theta,k}(v) = x^{\theta}(v)$ for all $k \geq n-1$,

$x^{\theta,1}(v) = b^v - g^v(N)e^i$ for $i \in N$ satisfying $\theta(i) = 1$.

(iii) If $i \in N$, $1 \leq i < k$ and $S = \{\theta^{-1}(j) \mid 1 \leq j \leq i\}$, then $\sum_{j \in S} x_j^{\theta,k}(v) = v(S)$.

(iv) If (7.2) holds and $x^{\theta,k}(v) \in C(v)$, then $x^{\theta,k}(v) \in \text{ext } C(v)$.

PROOF. (i) Let $i \in N$. By applying (5.10) to the game $v_k$, we have

$$x_i^{\theta}(v_k) = v_k(P_i^{\theta} \cup \{i\}) - v_k(P_i^{\theta}).$$

If $\theta(i) < k$, then $|P_i^{\theta}| < k-1$ and hence, it follows that

$$x_i^{\theta}(v_k) = v_k(P_i^{\theta} \cup \{i\}) - v_k(P_i^{\theta})$$
$$= v(P_i^{\theta} \cup \{i\}) - v(P_i^{\theta}) = x_i^{\theta,k}(v).$$

If $\theta(i) > k$, then $|P_i^{\theta}| \geq k$ and hence,

$$x_i^{\theta}(v_k) = b^v(P_i^{\theta} \cup \{i\}) - g^v(N) - b^v(P_i^{\theta}) + g^v(N)$$
$$= b_i^v = x_i^{\theta,k}(v).$$

If $\theta(i) = k$, then $|P_i^{\theta}| = k-1$ and hence,

$$x_i^{\theta}(v_k) = b^v(P_i^{\theta} \cup \{i\}) - g^v(N) - v(P_i^{\theta})$$
$$= v(N) - v(P_i^{\theta}) - \sum_{j;\theta(j)>k} b_j^v = x_i^{\theta,k}(v).$$

In any case, $x_i^{\theta}(v_k) = x_i^{\theta,k}(v)$. So, (i) holds.

(ii) For any $k \geq n-1$ we get $v_k = v$. Together with part (i), this implies that $x^{\theta,k}(v) = x^{\theta}(v_k) = x^{\theta}(v)$ for all $k \geq n-1$. If $\theta(i) = 1$, then $P_i^{\theta} = \emptyset$ and therefore, it follows from the formula (7.24) applied to $k = 1$ that

$$x_i^{\theta,1}(v) = v(N) - b^v(N-\{i\}) = b_i^v - g^v(N),$$

$$x_j^{\theta,1}(v) = b_j^v \qquad \text{for all } j \in N-\{i\}.$$ So, (ii) holds.

(iii) Let $i \in N$, $1 \leq i < k$ and put $S := \{\theta^{-1}(j) \mid 1 \leq j \leq i\}$. Then we have $v_k(S) = v(S)$ since $|S| = i < k$. Now we obtain

$$\sum_{j\in S} x_j^{\theta,k}(v) = \sum_{j\in S} x_j^{\theta}(v_k) = v_k(S) = v(S)$$

where the first equality follows from part (i) and the second equality from Proposition V.3.4(i) applied to the game $v_k$. So, (iii) holds.

(iv) Suppose that (7.2) holds. Then $C(v) = C(v_k)$ because of Lemma 1.4(iv). In view of this and part (i), it suffices to establish that $x^{\theta}(v_k) \in C(v_k)$ implies $x^{\theta}(v_k) \in \text{ext } C(v_k)$. However, the above implication is precisely the statement in Proposition V.3.4(iii) applied to the game $v_k$. So, (iv) holds. □

According to Lemma 4.2(iii), any adjusted marginal worth vector with respect to a natural number k in an n-person game is a payoff vector that satisfies at least k coalition constraints for the core of the given game, including the efficiency principle. In case (7.2) holds, then an adjusted marginal worth vector is either a point outside the core or a vertex of the core. Although the adjusted marginal worth vectors are in general not closely related to the core of the game, the convex hull of the set of all $((n-k)!)^{-1}n!$ adjusted marginal worth vectors is always a core catcher.

THEOREM 4.3. $C(v) \subset \text{conv } \{x^{\theta,k}(v) \mid \theta \in \Theta^n\}$ for all $v \in G^n$ and all $k \in \mathbb{N}$.

The next fundamental theorem concerning k-convexity deals with three characterizations of k-convex games in terms of the adjusted marginal worth vectors and the core.

A first characterization expresses that an n-person game v is k-convex if and only if the inclusion of Theorem 4.3 is an equality, i.e., the core catcher in question coincides with the core C(v) of the game v. In other words, for the k-convexity of an n-person game it is both necessary and sufficient that all $((n-k)!)^{-1}n!$ corresponding adjusted marginal worth vectors belong to the core of the game. A third characterization states that a game is k-convex if and only if the vertices of the core of the given game are precisely the adjusted marginal worth vectors.

THEOREM 4.4. Let $v \in G^n$ and $k \in \mathbb{N}$. The following four statements are equivalent.

(i) v is a k-convex game.

(ii) $x^{\theta,k}(v) \in C(v)$ for all $\theta \in \Theta^n$.

(iii) $C(v) = \text{conv}\,\{x^{\theta,k}(v) \mid \theta \in \Theta^n\}$.

(iv) $\text{ext}\,C(v) = \{x^{\theta,k}(v) \mid \theta \in \Theta^n\}$.

PROOF of Theorem 4.3.
Let $v \in G^n$ and $k \in \mathbb{N}$. Then $C(v) \subset C(v_k)$ by Lemma 1.4(iii). Further, $x^{\theta}(v_k) = x^{\theta,k}(v)$ for all $\theta \in \Theta^n$ by Lemma 4.2(i). Now we obtain the inclusions

$$C(v) \subset C(v_k) \subset \text{conv}\{x^{\theta}(v_k) \mid \theta \in \Theta^n\} = \text{conv}\{x^{\theta,k}(v) \mid \theta \in \Theta^n\}$$

where the second inclusion follows from Theorem V.3.6 applied to the game $v_k$. This completes the proof. □

PROOF of Theorem 4.4.
We prove (i) ⇒ (iv) ⇒ (iii) ⇒ (ii) ⇒ (i). The implication (iii) ⇒ (ii) trivially holds. By Lemma 4.2(i), we always have that $x^{\theta,k}(v) = x^{\theta}(v_k)$ for all $\theta \in \Theta^n$.

(a) Suppose that (i) holds. By the k-convexity of v, both the game $v_k$ is convex and (7.2) holds. Then $C(v) = C(v_k)$ because of Lemma 1.4(iv). Now we obtain the equalities

$$\text{ext}\,C(v) = \text{ext}\,C(v_k) = \{x^{\theta}(v_k) \mid \theta \in \Theta^n\} = \{x^{\theta,k}(v) \mid \theta \in \Theta^n\}$$

where the second equality follows from Theorem V.3.7(iv) applied to the convex game $v_k$. So, (i) implies (iv).

(b) Suppose that (iv) holds. In particular, $x^{\theta,k}(v) \in C(v)$ for all $\theta \in \Theta^n$. Since the core is a convex set, we obtain

$$\text{conv}\,\{x^{\theta,k}(v) \mid \theta \in \Theta^n\} \subset C(v).$$

The inverse inclusion is also valid because of Theorem 4.3. Hence, the equality in (iii) holds. So, (iv) implies (iii).
(c) Suppose that (ii) holds. We first establish (7.2). Let $S \subset N$ with $|S| \geq k$. Let $\theta: N \to N$ be any permutation such that $S = \{\theta^{-1}(j) \mid 1 \leq j \leq |S|\}$. We have that $\theta(j) > |S| \geq k$ for all $j \in N-S$ and thus, $x_j^{\theta,k}(v) = b_j^v$ for all $j \in N-S$. Now we obtain

$$g^v(N) = b^v(N) - v(N) = b^v(S) + \sum_{j\in N-S} x_j^{\theta,k}(v) - v(N)$$

$$= b^v(S) - \sum_{j\in S} x_j^{\theta,k}(v) \leq b^v(S) - v(S) = g^v(S)$$

where the inequality is due to $x^{\theta,k}(v) \in C(v)$. Therefore, (7.2) holds. Then $C(v) = C(v_k)$ because of Lemma 1.4(iv). By using part (ii), it follows that

$$x^\theta(v_k) = x^{\theta,k}(v) \in C(v) = C(v_k) \qquad \text{for all } \theta \in \Theta^n.$$

From this and Theorem V.3.7(i)-(ii) applied to the game $v_k$, we conclude that $v_k$ is a convex game. Hence, the game v is k-convex. So, (ii) implies (i). □

Concerning the characterizations of convex as well as k-convex n-person games, we emphasize that Theorem V.3.7 involving convexity can be regarded as the special case of Theorem 4.4 applied to k-convexity where $k = n$. The above proof of Theorem 4.4 is mainly based on Theorem V.3.7. An alternative proof of Theorem 4.4 which does not make use of the fundamental Theorem V.3.7 concerning convexity, was presented in Driessen (1986b). We elucidate the results of Theorem 4.4 by means of two examples.

EXAMPLE 4.5. Consider again the 4-person game v of Example 3.7. Then we have $v(N) = 5$, $b_i = 2$ for all $i \in N$, $g(N) = 3$ and $g(S) = 2|S|$ if $|S| \leq 2$. We first note that the game v is not convex since the corresponding gap function g is not monotonic and moreover, the game v is not 1-convex because of $g(N) > g(\{i\})$ for all $i \in N$. According to Theorem 2.2,

for the 2-convexity of the 4-person game v it is necessary and sufficient that the following conditions hold:
$g(\{i,j\}) \geq g(N)$ and $g(\{i\}) + g(\{j\}) \geq g(N) \geq g(\{i\})$ for all $i,j \in N$, $i \neq j$. Obviously, the game v is 2-convex. Now it follows from Theorem 4.4(iv) that the extreme points of the core of the game are precisely the twelve adjusted marginal worth vectors with respect to the natural number $k = 2$. Each extreme point of the core has multiplicity 2 and these twelve points are given by

$$\begin{array}{lll} (0,1,2,2), & (0,2,1,2), & (0,2,2,1), \\ (1,0,2,2), & (2,0,1,2), & (2,0,2,1), \\ (1,2,0,2), & (2,1,0,2), & (2,2,0,1), \\ (1,2,2,0), & (2,1,2,0), & (2,2,1,0). \end{array}$$

EXAMPLE 4.6. Consider once again the n-person game v of Example 3.2 where $n \geq 3$ and $n > 2m-1$. Then we get $v(N) = n$, $b_i = 1$ for all $i \in N$ and further,

$$\begin{aligned} g(S) &= |S|-m+1 && \text{if } m \leq |S| < \tfrac{1}{2}(n+1) \\ &= 0 && \text{otherwise.} \end{aligned}$$

Thus, the gap function g is nonnegative such that $g(N) = 0$. Now we deduce from Proposition III.3.4 that the core of the game is a singleton consisting of the upper vector, i.e., $C(v) = \{b\} = \{1_n\}$. For any permutation $\theta$ on N we show that the corresponding adjusted marginal worth vector $x^{\theta,k}(v)$ belongs to the core of the game v if and only if $1 \leq k \leq m$. Let $\theta \in \Theta^n$. In case $k \in \mathbb{N}$ such that $k > m$, then $\theta(i) = m$ implies $|P_i^\theta| = m-1$ as well as $\theta(i) < k$ and hence,

$$x_i^{\theta,k}(v) = v(P_i^\theta \cup \{i\}) - v(P_i^\theta) = 0 \qquad \text{if } \theta(i) = m.$$

Particularly, it follows that $x^{\theta,k}(v) \notin C(v)$ when $k > m$. In case $1 \leq k \leq m$, then the formulas (7.24) and (7.10) yield

$$\begin{aligned} x_i^{\theta,k}(v) &= v(P_i^\theta \cup \{i\}) - v(P_i^\theta) = 1 && \text{if } \theta(i) < k \\ &= b_i = 1 && \text{if } \theta(i) > k \\ &= 1 && \text{if } \theta(i) = k. \end{aligned}$$

So, $x^{\theta,k}(v) = 1_n \in C(v)$ whenever $1 \leq k \leq m$. In view of Theorem 4.4(i)-(ii), we conclude that v is a k-convex game iff $1 \leq k \leq m$. Of course, this conclusion is in agreement with the result mentioned at the end of Example 3.8.

COROLLARY 4.7. Both the $\tau$-value $\tau(v)$ and the nucleolus $\eta(v)$ of a 1-convex n-person game v are the centre of gravity of the extreme points of the core of the game, i.e.,

$$\tau(v) = \eta(v) = b^v - n^{-1}g^v(N)1_n \qquad \text{for all } v \in C_1^n.$$

PROOF. Let $v \in C_1^n$. Then we conclude from Theorem 4.4(iv) applied to k = 1 and Lemma 4.2(ii) respectively that

$$\text{ext } C(v) = \{x^{\theta,1}(v) \mid \theta \in \Theta^n\} = \{b - g(N)e^i \mid i \in N\}.$$

Because each extreme point $b - g(N)e^i$, $i \in N$, of the core has multiplicity $(n-1)!$, we have that the centre of gravity of the extreme points of the core equals $b - n^{-1}g(N)1_n$. In view of Theorem III.5.3, it is now clear that the $\tau$-value of the 1-convex game v coincides with the centre of gravity of the core. Concerning the nucleolus, it is sufficient to prove that (4.10) holds because of Theorem IV.3.1(ii). Here the condition (4.10) in terms of savings is equivalent to the condition (4.9) in terms of costs. Evidently, (4.9) is a direct consequence of the 1-convexity condition (4.6) in terms of costs. Hence, (4.10) holds. □

The nucleolus of a 1-convex game v is given by the simple formula as mentioned in Corollary 4.7. However, an explicit formula for the nucleolus of a k-convex game where $k \geq 2$ is not available in spite of the known fact that the nucleolus occupies a central position within the regular core.

## 5. The $\tau$-value of a k-convex game

Due to Theorem III.5.3, the $\tau$-value payoff vector for a 1-convex n-person game v is determined by the egalitarian partition of the joint concession amount $g^v(N)$ of the grand coalition, i.e.,

$$\tau(v) = b^v - n^{-1}g^v(N)1_n \qquad \text{for all } v \in C_1^n.$$

We recall that the k-convexity of a game gives rise to a specific division of the gaps into three levels in such a way that the corresponding gap function is monotonic at the lower level which consists of gaps of coalitions with at most k-1 players. As a consequence, the gap function $g^v$ of a k-convex game v where $k \geq 2$ possesses the property that $0 \leq g^v(\{i\}) \leq g^v(S)$ whenever $i \in S$. Thus, k-convexity where

$k \geq 2$ implies semiconvexity and therefore, the $\tau$-value of a k-convex game where $k \geq 2$ can be determined with the aid of Theorem III.6.2. As such, the $\tau$-value payoff vector for a zero-normalized k-convex game where $k \geq 2$ is proportional to the upper vector of the game.

THEOREM 5.1. Let $k \in \mathbb{N}$ and $v \in C_k^n$.

(i) If $g^v(N) = 0$, then $\tau(v) = b^v$.

(ii) If $g^v(N) > 0$ and $k = 1$, then $\tau(v) = b^v - n^{-1}g^v(N)1_n$.

(iii) If $g^v(N) > 0$ and $k \geq 2$, then

$$\tau(v) = b^v - g^v(N)\Big(\sum_{j\in N} g^v(\{j\})\Big)^{-1}\big(g^v(\{1\}),\ldots,g^v(\{n\})\big).$$

(iv) If $g^v(N) > 0$, $k \geq 2$ and $v(\{i\}) = 0$ for all $i \in N$, then $\tau(v) = v(N)\big(b^v(N)\big)^{-1}b^v$.

The $\tau$-value of a k-convex game does not necessarily belong to the core of the game. However, for certain small numbers of k or n it is guaranteed that the $\tau$-value of a k-convex n-person game lies inside the core of the game.

THEOREM 5.2. Let $k \in \mathbb{N}$ and $v \in C_k^n$. If $k \leq 2$ or $n \leq 4$, then $\tau(v) \in C(v)$.

PROOF. (a) Put $x := \tau(v)$. Then $x(N) = v(N)$ and $x_i \leq b_i$ for all $i \in N$ by the definition of the $\tau$-value. Now it follows from Proposition 1.2 that $x \in C(v)$ if and only if

$$x(S) \geq v(S) \qquad \text{for all } S \neq \emptyset \text{ satisfying } g(S) < g(N).$$

Let $S \subset N$, $S \neq \emptyset$, be such that $g(S) < g(N)$. Because (7.2) holds by the k-convexity of v, we obtain $1 \leq |S| \leq k-1$. In case $k = 1$, then all coalition constraints for the core are already satisfied by the pre-imputation x. In case $k = 2$, then the remaining constraints $x_i \geq v(\{i\})$ for all $i \in N$ are valid because the $\tau$-value is individually rational. So, $\tau(v) \in C(v)$ whenever $k \leq 2$.

(b) In case $n \leq 3$, then $\tau(v) \in C(v)$ by Corollary III.3.6. It remains to consider the case $n = 4$, $k \geq 3$. By $k \geq n-1$, the notion of k-convexity is equivalent to convexity and hence, the k-convex 4-person game v is zero-monotonic as well as semiconvex. Now we conclude from Theorem III.6.3 that $\tau(v) \in C(v)$. This completes the proof. □

By Corollary 4.7, the $\tau$-value of a 1-convex game coincides with both the nucleolus and the centre of gravity of the core of the game. Although the $\tau$-value of a 2-convex game is included in the core, it may differ from the nucleolus and the centre of gravity of the core as illustrated by the next example.

EXAMPLE 5.3. Let the 4-person game v be given by

$$\begin{array}{llll} v(i) = 0 \text{ for all } i \in N, & & v(N) = 20, \\ v(12) = v(13) = 1, & & v(14) = 3, \\ v(23) = 4, & v(24) = 6, & v(34) = 7, \\ v(123) = 9, & & v(124) = 12, \\ v(134) = 13, & & v(234) = 15. \end{array}$$

Since the upper vector $b = (5,7,8,11)$, the corresponding adjusted marginal worth vectors with respect to the natural number $k = 2$ are as follows:

$$\begin{array}{lll} (0,1,8,11), & (0,7,2,11), & (0,7,8,5), \\ (1,0,8,11), & (5,0,4,11), & (5,0,8,7), \\ (2,7,0,11), & (5,4,0,11), & (5,7,0,8), \\ (5,7,8,0). & & \end{array}$$

Each vector has multiplicity 2 with the exception of the vector (5,7,8,0) which is associated with six orderings of the player set N. It is straightforward to verify that the ten above adjusted marginal worth vectors belong to the core of the game v and so, the 4-person game v is 2-convex by means of Theorem 4.4(i)-(ii). By taking into account the multiplicity of the above extreme points of the core, we obtain that the centre of gravity of the core of the game v is equal to $\frac{1}{6}(19,27,31,43)$. From Theorem 5.1(iv) we deduce that the $\tau$-value payoff vector for the 2-convex game v is given by $\tau(v) = \alpha b$ where $\alpha = 20/31$. Here it is left to the reader to verify that the nucleolus $\eta(v) = \frac{1}{6}(15,25,31,49)$. We observe that the $\tau$-value, the nucleolus and the centre of gravity of the core are three different points inside the core of the 2-convex 4-person game v. In addition, another core-element is given by the Shapley value payoff vector $\phi(v) = \frac{1}{12}(37,57,63,83)$.

## 6. The Shapley value of a k-convex game

According to Theorem V.4.2, the Shapley value of a convex game is the centre of gravity of the extreme points of the core of the game where a certain multiplicity of the marginal worth vectors is taken into account.
For a k-convex n-person game where $1 \le k \le n-2$ the Shapley value is in general not closely related to the centre of gravity of the extreme points of the core since the Shapley value is based on the worths of all coalitions, whereas the adjusted marginal worth vectors with respect to the natural number k do not depend upon the worths of coalitions with at least k players and at most n-2 players. Nevertheless, the Shapley value of a k-convex n-person game can still be described in terms of the centre of gravity of the core and a certain correction-vector when the game is symmetric for coalitions with at least k players and at most n-2 players.

THEOREM 6.1. Let $k \in \mathbb{N}$, $v \in C_k^n$ be such that $1 \le k \le n-1$ and

$$v(S) = v(T) \qquad \text{for all} \quad S, T \subset N \tag{7.25}$$

such that $|S| = |T|$ and $k \le |S| \le n-2$.
Then

$$\phi(v) = (n!)^{-1} \sum_{\theta\in\Theta^n} x^{\theta,k}(v)$$

$$+ \big(n(n-1)\big)^{-1}(n-1-k)\big(b^v(N)1_n - nb^v\big).$$

PROOF. By the formula (5.13), we have $n!\phi(v) = \sum_{\theta\in\Theta^n} x^\theta(v)$.
Thus, the statement in the theorem reduces to the equality

$$\sum_{\theta\in\Theta^n} \big(x^\theta(v) - x^{\theta,k}(v)\big) = (n-2)!(n-1-k)\big(b(N)1_n - nb\big). \tag{7.26}$$

In case $k = n-1$, then $x^{\theta,k}(v) = x^\theta(v)$ for all $\theta \in \Theta^n$ by using Lemma 4.2(ii). Therefore, (7.26) trivially holds in case $k = n-1$. It remains to consider the case $1 \le k \le n-2$. Let $i \in N$. Given any ordering $\theta \in \Theta^n$, any player $j \in N$ and any natural number $s \in \mathbb{N}$ satisfying $k \le s \le n-2$, we define the vector $y^\theta \in \mathbb{R}^n$, the set $V_j$ of permutations on N and the real number $\alpha_s$ respectively by

$$y^\theta := x^\theta(v) - x^{\theta,k}(v),$$

$$V_j := \{\sigma \in \Theta^n \mid \sigma(i) = j\},$$

$$\alpha_s := v(S) \qquad \text{for all } S \subset N \text{ with } |S| = s.$$

The real numbers $\alpha_s$, $k \le s \le n-2$, are well-defined because of the symmetry condition (7.25) for the game v. Evidently, $|V_j| = (n-1)!$ for all $j \in N$. In view of the formula (5.10), the i-th coordinate of the vector $y^\theta$ is given by

$$y_i^\theta = v(P_i^\theta \cup \{i\}) - v(P_i^\theta) - x_i^{\theta,k}(v) \qquad \text{for all } \theta \in \Theta^n.$$

(a) If $1 \le \theta(i) < k$, then it is a direct consequence of the formula (7.24) that

$$y_i^\theta = v(P_i^\theta \cup \{i\}) - v(P_i^\theta) - x_i^{\theta,k}(v) = 0.$$

(b) If $\theta(i) = n$, then $P_i^\theta = N-\{i\}$, $\theta(i) > k$, and so,

$$y_i^\theta = v(N) - v(N-\{i\}) - x_i^{\theta,k}(v) = b_i - x_i^{\theta,k}(v) = 0.$$

(c) If $k < \theta(i) \le n-2$, say $\theta(i) = m$, then $|P_i^\theta| = m-1$ and so,

$$y_i^\theta = v(P_i^\theta \cup \{i\}) - v(P_i^\theta) - b_i = \alpha_m - \alpha_{m-1} - b_i. \quad \text{Hence,}$$

$$\sum_{m=k+1}^{n-2} \sum_{\theta \in V_m} y_i^\theta = \sum_{m=k+1}^{n-2} \sum_{\theta \in V_m} \left(\alpha_m - \alpha_{m-1} - b_i\right)$$

$$= (n-1)! \sum_{m=k+1}^{n-2} \left(\alpha_m - \alpha_{m-1} - b_i\right)$$

$$= (n-1)!(\alpha_{n-2} - \alpha_k) - (n-1)!(n-2-k)b_i.$$

(d) If $\theta(i) = n-1$, then $|P_i^\theta| = n-2$, $\theta(i) > k$, and so,

$$y_i^\theta = v(P_i^\theta \cup \{i\}) - v(P_i^\theta) - b_i$$

$$= v(P_i^\theta \cup \{i\}) - \alpha_{n-2} - b_i. \qquad \text{Hence,}$$

$$\sum_{\theta \in V_{n-1}} y_i^\theta = \sum_{\theta \in V_{n-1}} v(P_i^\theta \cup \{i\}) - (n-1)!\alpha_{n-2} - (n-1)!b_i.$$

(e) If $\theta(i) = k$, then $|P_i^\theta \cup \{i\}| = k$ and so,

$$y_i^\theta = v(P_i^\theta \cup \{i\}) - v(P_i^\theta) - v(N) + v(P_i^\theta) + \sum_{j;\theta(j)>k} b_j$$

$$= \alpha_k - v(N) + b(N) - b_i - b(P_i^\theta). \qquad \text{Hence,}$$

$$\sum_{\theta\in V_k} y_i^\theta = (n-1)!\left(\alpha_k - v(N) + b(N) - b_i\right) - \sum_{\theta\in V_k} b(P_i^\theta).$$

(f) Summing up the relevant equalities in the parts (a)-(e) over all permutations on N, we obtain

$$\sum_{\theta\in\Theta^n} y_i^\theta = \sum_{\theta\in V_{n-1}} v(P_i^\theta \cup \{i\}) - \sum_{\theta\in V_k} b(P_i^\theta)$$

$$- (n-1)!v(N) + (n-1)!b(N) - (n-1)!(n-k)b_i.$$

Straightforward combinatorial arguments yield

$$\sum_{\theta\in V_k} b(P_i^\theta) = (n-2)!(k-1)b(N-\{i\}) \qquad \text{and}$$

$$\sum_{\theta\in V_{n-1}} v(P_i^\theta \cup \{i\}) = (n-2)! \sum_{j\in N-\{i\}} v(N-\{j\})$$

$$= (n-2)! \sum_{j\in N-\{i\}} \left(v(N) - b_j\right)$$

$$= (n-1)!v(N) - (n-2)!b(N-\{i\}).$$

Now it follows that

$$\sum_{\theta\in\Theta^n} \left(x_i^\theta(v) - x_i^{\theta,k}(v)\right) = \sum_{\theta\in\Theta^n} y_i^\theta$$

$$= -(n-2)!kb(N-\{i\}) + (n-1)!b(N) - (n-1)!(n-k)b_i$$

$$= (n-2)!(n-1-k)b(N) + (n-2)!\left[k - (n-1)(n-k)\right]b_i$$

$$= (n-2)!(n-1-k)\left(b(N) - nb_i\right). \qquad \text{So, (7.26) holds.} \quad \square$$

The above theorem expresses that, if the symmetry condition (7.25) is satisfied by a k-convex n-person game v where $1 \le k \le n-1$, then the Shapley value $\phi(v)$ of the game v is equal to the centre of gravity of the extreme points of the core (where a certain multiplicity is taken into account) added to a special correction-vector. It is obvious that the correction-vector in question vanishes if and only if

$$k = n-1 \qquad \text{or} \qquad b_i^v = b_j^v \text{ for all } i,j \in N.$$

The latter condition requires that the n-person game v is symmetric for the (n-1)-person coalitions. In case $k = n-1$, then the symmetry condition (7.25) is superfluous, while k-convexity and convexity are equivalent notions for the n-person game v. So, Theorem 6.1 applied to $k = n-1$ reduces to the known case for a convex n-person game.

EXAMPLE 6.2. Consider again the 1-convex 3-person game v of Example III.5.6. The game satisfies the symmetry condition (7.25) because $v(\{i\}) = 0$ for all $i \in N$. From Figure 3.2 we deduce that the centre of gravity of the extreme points of the core of the game is equal to the payoff vector (2,8,8). Furthermore, the correction-vector in question is given by (2,-1,-1). Now it follows from Theorem 6.1 that the Shapley value $\phi(v) = (4,7,7)$. We observe that the Shapley value of the 1-convex game v falls outside the core in spite of the fact that the symmetry condition (7.25) holds.

## 7. The kernel and the prekernel of a k-convex game

According to Theorem V.7.3, zero-monotonicity (and in particular, convexity) is a sufficient condition for the kernel to coincide with the prekernel. However, k-convexity does not imply zero-monotonicity and consequently, the kernel of a k-convex game may differ from the prekernel as illustrated by the next example.

EXAMPLE 7.1. Consider once again the 8-person game v of Example II.6.11 where $\alpha$ is any nonnegative real number. The (pre)kernel of the game v only contains pre-imputations of the form

$$x = x(\beta) = (\beta,\ \beta,\ \beta,\ \beta,\ 10-\beta,\ 10-\beta,\ 10-\beta,\ 10-\beta)$$

where $\beta \in \mathbb{R}$. Clearly, $x(\beta) \in I(v)$ iff $0 \le \beta \le 10$. Further, by (2.8) we obtain the equivalence

$$x(\beta) \in C(v) \qquad \text{iff} \qquad 4 \le \beta \le 6 \text{ and } \tfrac{\alpha}{2}-5 \le \beta \le 15-\tfrac{\alpha}{2}.$$

We look at the (pre)kernel of the game v where $14 < \alpha < 19$. By Example II.6.11, we get the following results concerning the kernel and the prekernel of the game v.

$$\mathcal{K}^*(v) = \{x(\beta) \mid 5-\tfrac{\alpha}{2} \le \beta \le \alpha-14 \text{ or}$$
$$\beta = 5 \text{ or } 24-\alpha \le \beta \le 5+\tfrac{\alpha}{2}\},$$

$$\mathcal{K}(v) = \{x(\beta) \mid 0 \le \beta \le \alpha-14 \text{ or}$$
$$\beta = 5 \text{ or } 24-\alpha \le \beta \le 10\}.$$

Thus, the prekernel consists of two disconnected straight line segments and a single point in such a way that a part of each line segment lies outside the imputation set I(v). Because the kernel is the intersection of the prekernel with the imputation set, it follows that the kernel of v is strictly included in the prekernel whenever $14 < \alpha < 19$. In view of Example 2.3, the game v is 2-convex iff $\alpha \le 16$. We conclude that the kernel of the 2-convex 8-person game v where $14 < \alpha \le 16$ differs from the prekernel.
In addition, we are interested in the geometric position of the (pre)kernel of the 2-convex game v with respect to the core C(v). In case $\alpha \le 18$, then we obtain the equivalence $x(\beta) \in C(v)$ iff $4 \le \beta \le 6$. Hence, the (pre)kernel of the 2-convex 8-person game v where $14 < \alpha \le 16$ consists of two disconnected straight line segments outside the core and only one point inside the core. It is left to the reader to verify that the core of the 2-convex 8-person game v has 56 extreme points.

From the above example we learn that the (pre)kernel of a k-convex game may contain points outside the core of the game. The next theorem expresses that the part of the (pre)kernel inside the core of a k-convex game consists of a unique point and as such, the single point coincides with the nucleolus of the k-convex game.

THEOREM 7.2. Let $k \in \mathbb{N}$. Then

$$\mathcal{K}(v) \cap C(v) = \mathcal{K}^*(v) \cap C(v) = \{\eta(v)\} \qquad \text{for all } v \in C_k^n.$$

PROOF. By Theorem II.6.4 applied to $\epsilon = 0$, the equality $\mathcal{K}(v) \cap C(v) = \mathcal{K}^*(v) \cap C(v)$ holds for all $v \in G^n$. Further, $\eta(v) \in \mathcal{K}(v) \cap C(v)$ for all $v \in G^n$ with $C(v) \ne \emptyset$ by the two Theorems II.7.8-7.9. Hence, it is sufficient to prove that the intersection $\mathcal{K}(v) \cap C(v)$ is a singleton for all $v \in C_k^n$.
Let $v \in C_k^n$. By the k-convexity of v, the game $v_k$ is convex and (7.2) holds. Then we get $C(v) = C(v_k)$ by Lemma 1.4(iv).

Now it follows from Corollary II.6.8 applied to the two games v and $v_k$ that $\mathcal{K}(v) \cap C(v) = \mathcal{K}(v_k) \cap C(v_k)$. Moreover, we deduce from Theorem V.7.4 applied to the convex game $v_k$ that $\mathcal{K}(v_k) = \{\eta(v_k)\} \subset C(v_k)$. We conclude that

$$\mathcal{K}(v) \cap C(v) = \mathcal{K}(v_k) \cap C(v_k) = \{\eta(v_k)\}.$$

Thus, the intersection of the kernel with the core of the k-convex game v consists of a unique point and as such, the intersection coincides with the nucleolus $\eta(v)$ of the given game v. So, $\mathcal{K}(v) \cap C(v) = \{\eta(v_k)\} = \{\eta(v)\}$. □

COROLLARY 7.3. If a game $v \in G^n$ is 1-convex, then

$$\mathcal{K}(v) \cap C(v) = \mathcal{K}^*(v) \cap C(v) = \{b^v - n^{-1}g^v(N)1_n\}.$$

## 8. The bargaining set $\mathcal{M}$ of a k-convex game

According to Theorem V.6.3, the bargaining set $\mathcal{M}$ of a convex game coincides with the core of the game. Because the bargaining set $\mathcal{M}$ always contains the core as well as the kernel, we learn from Example 7.1 that the bargaining set $\mathcal{M}$ of a k-convex n-person game where $1 \le k \le n-2$ does not necessarily coincide with the core of the game.

EXAMPLE 8.1. Consider again the 8-person game v of Example 7.1. We look at the kernel and the core of the game v where $14 < \alpha < 19$. Then we have

$$\mathcal{K}(v) = \{x(\beta) \mid 0 \le \beta \le \alpha-14 \text{ or } \beta = 5 \text{ or } 24-\alpha \le \beta \le 10\},$$

$$x(\beta) \in C(v) \text{ iff } 4 \le \beta \le 6 \qquad \text{in case } 14 < \alpha < 18,$$

$$x(\beta) \in C(v) \text{ iff } \tfrac{\alpha}{2}-5 \le \beta \le 15-\tfrac{\alpha}{2} \qquad \text{in case } 18 \le \alpha < 19.$$

$$\text{So, } \mathcal{K}(v) - C(v) = \{x(\beta) \mid 0 \le \beta \le \alpha-14 \text{ or } 24-\alpha \le \beta \le 10\} \quad \text{in case } 14 < \alpha < 18,$$

$$= \{x(\beta) \mid 0 \le \beta < \tfrac{\alpha}{2}-5 \text{ or } 15-\tfrac{\alpha}{2} < \beta \le 10\} \quad \text{in case } 18 \le \alpha < 19.$$

We obtain that the part of the kernel outside the core is nonempty and hence, the bargaining set $\mathcal{M}(v)$ of the game v strictly includes the core whenever $14 < \alpha < 19$. In view of Example 2.3, the game v is 2-convex iff $\alpha \le 16$. We conclude that the bargaining set $\mathcal{M}(v)$ of the 2-convex 8-person game v where $14 < \alpha \leq 16$ does not coincide with the core C(v) of the game v.

We also observe that the nonempty part of the bargaining set $\mathcal{M}(v)$ (and in particular, the kernel) outside the core of the game v is not a closed set whenever $18 \leq \alpha < 19$. According to the next theorem, this is due to the fact that the game v where $18 \leq \alpha < 19$ does not satisfy a certain condition which is closely related to the condition (7.2) applied to k = 2.

THEOREM 8.2. Let $v \in G^n$ be such that $C(v) \neq \emptyset$ and

$$g^v(S) > g^v(N) \text{ for all } S \subset N \text{ with } 2 \leq |S| \leq n-2. \quad (7.27)$$

If $\mathcal{M}(v) \neq C(v)$, then $\mathcal{M}(v) - C(v)$ is a nonempty closed set.

PROOF. Suppose $\mathcal{M}(v) \neq C(v)$. To prove the closedness of the set $\mathcal{M}(v) - C(v)$, let $x^{(m)}$, $m \in \mathbb{N}$, be a convergent sequence in $\mathcal{M}(v) - C(v)$. We show that the limit x of the sequence in question satisfies $x \in \mathcal{M}(v) - C(v)$. Clearly, $x \in \mathcal{M}(v)$ since the bargaining set $\mathcal{M}(v)$ itself is always closed.

It remains to show $x \notin C(v)$. Assume $x \in C(v)$. Then we have that $x_i \leq b_i$ for all $i \in N$ by Lemma III.1.2(i). From this we deduce that for all $S \subset N$

$$\begin{aligned} e(S,x) &= v(S) - x(S) = (b - x)(S) - g(S) \\ &\leq (b - x)(N) - g(S) = b(N) - v(N) - g(S) \\ &= g(N) - g(S). \end{aligned}$$

In view of (7.27), we obtain that $e(S,x) < 0$ for all $S \subset N$ with $2 \leq |S| \leq n-2$. Since the payoff vector x is the limit of the sequence $x^{(m)}$, $m \in \mathbb{N}$, it follows that

$$e(S,x^{(m)}) < 0 \qquad \text{for all } S \subset N \text{ with } 2 \leq |S| \leq n-2$$

whenever the number $m \in \mathbb{N}$ is large enough. From this and $x^{(m)} \in I(v) - C(v)$, we derive that there exists $j \in N$ with

$$e(N-\{j\},x^{(m)}) > 0 \qquad \text{or equivalently,} \qquad x_j^{(m)} > b_j.$$

From the positivity of the excess $e(N-\{j\},x^{(m)})$ we conclude that any player i, $i \neq j$, can object against player j with respect to the imputation $x^{(m)}$ by using the coalition $N-\{j\}$.

By Lemma III.1.2(ii), we have that $g(S) \geq 0$ for all $S \subset N$. In particular, $g(N) \geq 0$ and thus, $b(N) \geq v(N) = x^{(m)}(N)$. Together with the strict inequality $x_j^{(m)} > b_j$, this implies that there exists $i \in N$ with $x_i^{(m)} < b_i$. Now it follows that

$$e(T,x^{(m)}) < 0 \qquad \text{for all } T \in \Gamma_{ji} \text{ with } 2 \leq |T| \leq n-2,$$

$$e(\{j\},x^{(m)}) = v(\{j\}) - x_j^{(m)} < v(\{j\}) - b_j = -g(\{j\}) \leq 0,$$

$$e(N-\{i\},x^{(m)}) = v(N-\{i\}) - x^{(m)}(N-\{i\}) = x_i^{(m)} - b_i < 0.$$

So, $e(T,x^{(m)}) < 0$ for all $T \in \Gamma_{ji}$ and hence, there exists no counterobjection to any objection of player i against player j with respect to the imputation $x^{(m)}$. However, this is in contradiction with the fact $x^{(m)} \in \mathcal{M}(v)$. Now we have shown $x \notin C(v)$ which completes the proof of the closedness of the set $\mathcal{M}(v) - C(v)$. □

COROLLARY 8.3. Let $v \in C_1^n \cup C_2^n$ be such that (7.27) holds. If $\mathcal{M}(v) \neq C(v)$, then $\mathcal{M}(v) - C(v)$ is a nonempty closed set.

The corollary is a direct consequence of Theorem 8.2 since both 1-convex and 2-convex games possess a nonempty core. Moreover, it is not possible to state a similar corollary for k-convex n-person games where $3 \leq k \leq n-1$ because the k-convexity condition (7.3) is in contradiction with the condition (7.27) whenever $3 \leq k \leq n-1$.

Consider a balanced game v satisfying (7.27). In case the bargaining set $\mathcal{M}(v)$ does not coincide with the core $C(v)$, then the bargaining set $\mathcal{M}(v)$ can be partitioned into the two closed subsets $C(v)$ and $\mathcal{M}(v) - C(v)$. In topological terms, the bargaining set $\mathcal{M}(v)$ itself is a disconnected set whenever it differs from the core of the game v.

In addition to the above research on the determination of the bargaining set $\mathcal{M}$ of a k-convex n-person game where $1 \leq k \leq n-2$, we pose an open problem:

The bargaining set $\mathcal{M}$ is the union of the core and the kernel, i.e., $\mathcal{M}(v) = C(v) \cup \mathcal{K}(v)$ for all $v \in C_k^n$.

In this last chapter we have revealed the main part of the theory based on k-convex games. For minor results concerning the k-convexity notion and related subjects, we refer to Driessen (1985b, 1986a,b,c, 1987a).

# REFERENCES

AUMANN, R.J. and M. MASCHLER (1964). The bargaining set for cooperative games. In: *Advances in Game Theory* (Eds. M. Dresher, L.S. Shapley and A.W. Tucker). Princeton University Press, Princeton, New Jersey, 443-476.

--- (1985). Game theoretic analysis of a bankruptcy problem from the Talmud. *J. Econom. Theory* **36**, 195-213.

AUMANN, R.J. and L.S. SHAPLEY (1974). Values of non-atomic games. Princeton University Press, Princeton, New Jersey.

BAKER, M.J. and ASSOCIATES (1965). Runway cost impact study. *Report presented to the Association of Local Transport Airlines*, Jackson, Mississippi.

BANZHAF, J.F., III (1965). Weighted voting doesn't work: a mathematical analysis. *Rutgers Law Rev.* **19**, 317-343.

BOLGER, E.M. (1980). A class of power indices for voting games. *Internat. J. Game Theory* **9**, 217-232.

--- (1982). Characterizing the Banzhaf and Shapley values assuming limited linearity. *Internat. J. Game Theory* **11**, 1-12.

BONDAREVA, O.N. (1963). Some applications of linear programming methods to the theory of cooperative games (in Russian). *Problemy Kibernet.* **10**, 119-139.

CHETTY, V.K., D. DASGUPTA and T.E.S. RAGHAVAN (1976). Power and distribution of profits. *Discussion Paper* no. **139**, Indian Statistical Institute, Delhi Centre, New Delhi.

COLEMAN, J.S. (1971). Control of collectivities and the power of a collectivity to act. In: *Social Choice* (Ed. B. Lieberman). Gordon and Breach, London, 269-300.

CURIEL, I.J. (1987). A class of non-normalized power indices for simple games. *Math. Social Sci.* **13**, 141-152.

CURIEL, I.J., M. MASCHLER and S.H. TIJS (1987). Bankruptcy games. *Z. Oper. Res. Ser.* **A 31**, 143-159.

DAVIS, M. and M. MASCHLER (1963). Existence of stable payoff configurations for cooperative games. *Bull. Amer. Math. Soc.* **69**, 106-108.

--- (1965). The kernel of a cooperative game. *Naval Res. Logist. Quart.* **12**, 223-259.

DEEGAN, J., Jr. and E.W. PACKEL (1978). A new index of power for simple n-person games. *Internat. J. Game Theory* **7**, 113–123.

--- (1983). To the (minimal winning) victors go the (equally divided) spoils: a new power index for simple n-person games. In: *Political and Related Models* (Eds. S.J. Brams, W.F. Lucas and P.D. Straffin, Jr.). Springer-Verlag, New York, 239–255.

DRIESSEN, T.S.H. (1985a). A new axiomatic characterization of the Shapley value. *Methods Oper. Res.* **50**, 505–517.

--- (1985b). Properties of 1-convex n-person games. *OR Spektrum* **7**, 19–26.

--- (1986a). Solution concepts of k-convex n-person games. *Internat. J. Game Theory* **15**, 201–229.

--- (1986b). k-Convex n-person games and their cores. *Z. Oper. Res. Ser.* **A 30**, 49–64.

--- (1986c). A note on a core catcher of a cooperative game. *Z. Oper. Res. Ser.* **A 30**, 239–241.

--- (1987a). The core of a cooperative game: bounds and characterizations. In: *Surveys in Game Theory and Related Topics* (Eds. H.J.M. Peters and O.J. Vrieze). CWI Tract 39, Centre for Mathematics and Computer Science, Amsterdam, 181–208.

--- (1987b). The $\tau$-value: a survey. In: *Surveys in Game Theory and Related Topics* (Eds. H.J.M. Peters and O.J. Vrieze). CWI Tract 39, Centre for Mathematics and Computer Science, Amsterdam, 209–213.

DRIESSEN, T.S.H. and S.H. TIJS (1983). The $\tau$-value, the nucleolus and the core for a subclass of games. *Methods Oper. Res.* **46**, 395–406.

--- (1984). Game-theoretic solutions for some economic situations. *Cahiers Centre Études Rech. Opér.* **26**, 51–58.

--- (1985a). The $\tau$-value, the core and semiconvex games. *Internat. J. Game Theory* **14**, 229–247.

--- (1985b). The cost gap method and other cost allocation methods for multipurpose water projects. *Water Resources Res.* **21**, 1469–1475.

DUBEY, P. (1975a). Some results on values of finite and infinite games. *Ph. D. Thesis*. Cornell University, Ithaca, New York.

--- (1975b). On the uniqueness of the Shapley value. *Internat. J. Game Theory* **4**, 131–139.

--- (1982). The Shapley value as aircraft landing fees - revisited. *Management Sci.* **28**, 869–874.

DUBEY, P. and L.S. SHAPLEY (1979). Mathematical properties of the Banzhaf power index. *Math. Oper. Res.* **4**, 99-131.

--- (1984). Totally balanced games arising from controlled programming problems. *Math. Programming* **29**, 245-267.

Federal Inter-Agency River Basin Committee (1950). Proposed practices for economic analysis of river basin projects. *Report*, Washington, D.C.

GILLIES, D.B. (1953). Some theorems on n-person games. *Ph. D. Thesis*. Princeton University Press, Princeton, New Jersey.

HEANEY, J.P. (1979). Efficiency/equity analysis of environmental problems - a game theoretic perspective. In: *Applied Game Theory* (Eds. S.J. Brams, A. Schotter and G. Schwödiauer). Physica-Verlag, Vienna, 352-369.

HEANEY, J.P. and R.E. DICKINSON (1982). Methods for apportioning the cost of a water resource project. *Water Resources Res.* **18**, 476-482.

ICHIISHI, T. (1981). Super-modularity: applications to convex games and to the greedy algorithm for LP. *J. Econom. Theory* **25**, 283-286.

Inter-Agency Committee on Water Resources (1958). Proposed practices for economic analysis of river basin projects. *Report*, Washington, D.C.

KALAI, E. and M. SMORODINSKY (1975). Other solutions to Nash's bargaining problem. *Econometrica* **43**, 513-518.

KALAI, E. and E. ZEMEL (1982a). Totally balanced games and games of flow. *Math. Oper. Res.* **7**, 476-478.

--- (1982b). Generalized network problems yielding totally balanced games. *Oper. Res.* **30**, 998-1008.

KOHLBERG, E. (1971). On the nucleolus of a characteristic function game. *SIAM J. Appl. Math.* **20**, 62-66.

LEGROS, P. (1984). Formation des coalitions et allocation des coûts: une approche par la théorie des jeux. *Ph. D. Thesis*. University of Paris XII, La Varenne St Hilaire.

--- (1986). Allocating joint costs by means of the nucleolus. *Internat. J. Game Theory* **15**, 109-119.

LITTLECHILD, S.C. and G. OWEN (1973). A simple expression for the Shapley value in a special case. *Management Sci.* **20**, 370-372.

LUCAS, W.F. (1968). A game with no solution. *Bull. Amer. Math. Soc.* **74**, 237-239.

--- (1969). The proof that a game may not have a solution. *Trans. Amer. Math. Soc.* **137**, 219-229.

--- (1983). Measuring power in weighted voting systems. In: *Political and Related Models* (Eds. S.J. Brams, W.F. Lucas and P.D. Straffin, Jr.). Springer-Verlag, New York, 183-238.

LUCAS, W.F. and M. RABIE (1982). Games with no solutions and empty cores. *Math. Oper. Res.* **7**, 491-500.

MASCHLER, M. (1966). The inequalities that determine the bargaining set $\mathcal{M}$. *Israel J. Math.* **4**, 127-134.

--- (1976). An advantage of the bargaining set over the core. *J. Econom. Theory* **13**, 184-192.

MASCHLER, M. and B. PELEG (1966). A characterization, existence proof and dimension bounds for the kernel of a game. *Pacific J. Math.* **18**, 289-328.

MASCHLER, M., B. PELEG and L.S. SHAPLEY (1972). The kernel and bargaining set for convex games. *Internat. J. Game Theory* **1**, 73-93.

--- (1979). Geometric properties of the kernel, nucleolus, and related solution concepts. *Math. Oper. Res.* **4**, 303-338.

O'NEILL, B. (1982). A problem of rights arbitration from the Talmud. *Math. Social Sci.* **2**, 345-371.

OWEN, G. (1975a). Multilinear extensions and the Banzhaf value. *Naval Res. Logist. Quart.* **22**, 741-750.

--- (1975b). On the core of linear production games. *Math. Programming* **9**, 358-370.

--- (1978a). Characterization of the Banzhaf-Coleman index. *SIAM J. Appl. Math.* **35**, 315-327.

--- (1978b). A note on the Banzhaf-Coleman axioms. In: *Game Theory and Political Science* (Ed. P.C. Ordeshook). New York University Press, New York, 451-461.

--- (1982). Game theory (second edition). Academic Press, New York.

PELEG, B. (1963). Existence theorem for the bargaining set $\mathcal{M}$. *Bull. Amer. Math. Soc.* **69**, 109-110.

--- (1967). Existence theorem for the bargaining set $\mathcal{M}$. In: *Essays in Mathematical Economics* (Ed. M. Shubik). Princeton University Press, Princeton, New Jersey, 53-56.

--- (1985). An axiomatization of the core of market games. *Publication RM*-**68**, Department of Mathematics, The Hebrew University, Jerusalem, Israel.

--- (1986). On the reduced game property and its converse. *Internat. J. Game Theory* **15**, 187-200.

RANSMEIER, J.S. (1942). The Tennessee Valley Authority: a case study in the economics of multiple purpose stream planning. The Vanderbilt University Press, Nashville, Tennessee.

ROSENMÜLLER, J. (1971). Kooperative Spiele und Märkte. Springer-Verlag, Berlin.

ROTH, A.E. (1976). Subsolutions and the supercore of cooperative games. *Math. Oper. Res.* **1**, 43-49.

--- (1977). A note on values and multilinear extensions. *Naval Res. Logist. Quart.* **24**, 517-520.

SCHMEIDLER, D. (1969). The nucleolus of a characteristic function game. *SIAM J. Appl. Math.* **17**, 1163-1170.

SHAPLEY, L.S. (1953). A value for n-person games. In: *Contributions to the Theory of Games II* (Eds. H. Kuhn and A.W. Tucker). Princeton University Press, Princeton, New Jersey, 307-317.

--- (1967). On balanced sets and cores. *Naval Res. Logist. Quart.* **14**, 453-460.

--- (1971). Cores of convex games. *Internat. J. Game Theory* **1**, 11-26.

SHAPLEY, L.S. and M. SHUBIK (1954). A method for evaluating the distribution of power in a committee system. *Amer. Political Sci. Rev.* **48**, 787-792.

--- (1963). The core of an economy with nonconvex preferences. *RM*-**3518**, The Rand Corporation, Santa Monica, CA.

--- (1966). Quasi-cores in a monetary economy with nonconvex preferences. *Econometrica* **34**, 805-827.

--- (1967). Ownership and the production function. *Quart. J. Econom.* **81**, 88-111.

--- (1969). On market games. *J. Econom. Theory* **1**, 9-25.

--- (1972). The assignment game I: the core. *Internat. J. Game Theory* **1**, 111-130.

SOBOLEV, A.I. (1973). The functional equations that give the payoffs of the players in an n-person game (in Russian). In: *Advances in Game Theory* (Ed. E. Vilkas). Izdat. "Mintis", Vilnius, 151-153.

--- (1975). The characterization of optimality principles in cooperative games by functional equations (in Russian). *Mathematical Methods in the Social Sciences* **6**, 94-151.

STRAFFIN, P.D., Jr. (1983). Power indices in politics. In: *Political and Related Models* (Eds. S.J. Brams, W.F. Lucas and P.D. Straffin, Jr.). Springer-Verlag, New York, 256-321.

STRAFFIN, P.D., Jr. and J.P. HEANEY (1981). Game theory and the Tennessee Valley Authority. *Internat. J. Game Theory* **10**, 35-43.

SUZUKI, M. and M. NAKAYAMA (1976). The cost assignment of the cooperative water resource development: a game theoretical approach. *Management Sci.* **22**, 1081-1086.

THOMPSON, G.F. (1971). Airport costs and pricing. *Unpublished Ph. D. Thesis*. University of Birmingham, England.

TIJS, S.H. (1981). Bounds for the core and the $\tau$-value. In: *Game Theory and Mathematical Economics* (Eds. O. Moeschlin and D. Pallaschke). North-Holland Publishing Company, Amsterdam, 123-132.

--- (1987). An axiomatization of the $\tau$-value. *Math. Social Sci.* **13**, 177-181.

TIJS, S.H. and T.S.H. DRIESSEN (1986a). Game theory and cost allocation problems. *Management Sci.* 32, 1015-1028.

--- (1986b). Extensions of solution concepts by means of multiplicative $\epsilon$-tax games. *Math. Social Sci.* **12**, 9-20.

--- (1987). The $\tau$-value as a feasible compromise between utopia and disagreement. In: *Axiomatics and Pragmatics of Conflict Analysis* (Eds. J.H.P. Paelinck and P.H. Vossen). Gower Publishing Company, Aldershot, England, 142-156.

TIJS, S.H. and F.A.S. LIPPERTS (1982). The hypercube and the core cover of n-person cooperative games. *Cahiers Centre Études Rech. Opér.* **24**, 27-37.

TIJS, S.H., T. PARTHASARATHY, J.A.M. POTTERS and V. RAJENDRA PRASAD (1984). Permutation games: another class of totally balanced games. *OR Spektrum* **6**, 119-123.

VON NEUMANN, J. (1928). Zur Theorie der Gesellschaftespiele. *Math. Ann.* **100**, 295-320.

VON NEUMANN, J. and O. MORGENSTERN (1944). Theory of games and economic behavior. Princeton University Press, Princeton, New Jersey.

Water Resources Council (1962). Policies, standards, and procedures in the formulation, evaluation, and review of plans for use and development of water and related land resources. *Senate Doc.* **97**, 87th Congress, 2nd Session.

WEBER, R.J. (1978). Probabilistic values for games. *Cowles Foundation Discussion Paper* no. **471R**, Yale University, New Haven, Connecticut.

YOUNG, H.P. (1985a). Monotonic solutions of cooperative games. *Internat. J. Game Theory* **14**, 65–72.

--- (1985b). Cost allocation: methods, principles, applications. North-Holland Publishing Company, Amsterdam.

--- (1987). On dividing an amount according to individual claims or liabilities. *Math. Oper. Res.* **12**, 398–414.

# AUTHOR INDEX

## SUBJECT INDEX

THEORY AND DECISION LIBRARY

SERIES C: GAME THEORY, MATHEMATICAL PROGRAMMING AND
OPERATIONS RESEARCH

*Already published:*

**Compromise, Negotiation and Group Decision**
*Edited by Bertrand R. Munier and Melvin F. Shakun*
ISBN 90-277-2625-6

**Models of Strategic Rationality**
*by Reinhard Selten*
ISBN 90-277-2663-9